Practical TIG (GTA) welding

Practical TIG (GTA) welding

A survey of the process and equipment

Peter W Muncaster

ABINGTON PUBLISHING

Woodhead Publishing Ltd in association with The Welding Institute
Cambridge England

Published by Abington Publishing,
Abington Hall, Abington,
Cambridge CB1 6AH, England

First published 1991, Abington Publishing

British Library Cataloguing in Publication Data
A catalogue record for this book is available from the British Library.

ISBN 1 85573 020 0

Designed by Andrew Jones (text) and Chris Feely (jacket).
Typeset by BookEns Ltd, Baldock, Herts.
Printed by St Edmundsbury Press, Bury St Edmunds, Suffolk.

Contents

Preface

Of all the welding processes tungsten inert gas (TIG) welding is one of the most versatile. Since its inception it has been improved and refined in terms of equipment, consumables and application and is capable of still further improvements.

The author intends this to be a practical book on TIG containing advice on such items as power sources, ancillary components, torches, gases, electrodes, jigs, fixtures and heat sinks, *etc*. In short, an extensive guide to its use. The book sets out to explain in clear language the various processes, equipment and associated terminology without reaching into the worlds of advanced electronics or metallurgy.

The author has been for many years at the sharp end of TIG application in industry and has endeavoured in the pages that follow to pass on as much as possible of the knowledge accumulated during that time.

Some of the information provided may be obvious to experienced welders. However it is hoped that even for experts some useful extra knowledge will be gained from reading the book which in most cases gives the author's considered opinions on the subject; not necessarily in agreement with other peoples' views.

Certain passages in this book have been quoted almost verbatim from literature supplied by some of the companies listed. This is not for any reason other than that the passages quoted are definitive and could not significantly be improved to any great extent, as the basic TIG welding principles still apply. Also some extra tabulated information has been included, as such data can often be difficult to find for someone not totally involved in a particular industry, *e.g.* pipeline welding and pipe sizes.

Technical data for this book have been supplied by

Contributor	Subject
The Welding Institute (now TWI)	General information
Gordon Poxon, Esq	Archive material
British Oxygen Ltd	Industrial gases and consumables
Murex Welding Products	Power sources
Oxford Products Ltd	Power sources
Precision Systems Ltd	Power sources and systems
Power Electronic Consultants Ltd	Control systems
Elmer Wallace Plansee	Electrodes and grinders
Air Products Ltd	Industrial gases
Nederman Ltd	Fume extraction equipment
ESAB Ltd	Equipment and consumables
Interlas Ltd	Power sources
Astro-Arc (USA)	Orbital welding equipment
Arc Machines Inc (USA)	Orbital welding equipment
Welding Torch Co Ltd	Torches and cables
Redman Controls Ltd	Seam followers, arc length controllers
Vernon Boyd, Esq	Advice on computer controls
Keith Armstrong, Esq	General pipe data
Dr Bill Lucas (TWI)	Advice and encouragement
MacGregor Welding Systems Ltd	Low current power sources
Camarc Welding Equipment Ltd	Wire feeders, etc
Hobart Bros (GB) Ltd	Power sources
T J Handlers (Chelmsford) Ltd	Advice and assistance
FERLIN (Holland)	Tungsten grinding machines

If any contributors have not been acknowledged the author offers his most sincere apologies.

An enormous vote of thanks must be given to Mrs Hazel Muncaster who spent many unpaid hours deciphering the author's handwritten sheets of text and converting them to typed sense and order on an old, portable typewriter.

Introduction

A brief history of the TIG welding process

It is very difficult to establish a date at which anything in engineering can be said to be truly invented. All electric fusion welding can be classified as arc welding but, as near as can be established, TIG welding originated in the USA, probably before 1939 where it was used mainly to facilitate the then rather difficult process of welding aluminium. As the less dense than air inert gas helium occurred naturally in the USA and was readily and cheaply available, it was used as an inert shield to prevent undue oxidation of the aluminium during melt down. It was known as the Heliarc process.

TIG welding appears to have become commercially available in the UK and Europe after development by BOC around 1948. However, helium was (and still is) very expensive in Europe so BOC turned to argon (an inert gas denser than air) which was produced by BOC as a by-product of air liquefaction for oxygen production. BOC continued to develop both gas and equipment and put on the market a commercial TIG welding process known as Argonarc; argon having certain advantages over helium, other than cost. In passing, it should be mentioned that both the Heliarc and Argonarc methods operated on AC at that time because the arc acted as an electrical rectifier. However, the DC component of the welding current caused dirty welds with oxides entrapped in the weld bead. To counteract this so called rectifier, manufacturers in the USA used an extremely high voltage with a high frequency (HF) spark injector. This system was only partially effective in cleaning the weld and caused considerable radio frequency interference (RFI).

BOC's method eliminated the DC component by using a suppressor, a large capacitor connected in series with the arc, which gave excellent results for DC welding of aluminium but still gave some RFI. Use of argon as a shielding gas gave many advantages over helium, not least of which was a reduction in arc voltage to below 20 V. Argon also gave a smoother and much less fierce arc and cost was about 30% the cost of helium.

BOC also developed a surge injection system for arc starting and maintenance in which the original spark injection was used in short bursts for starting only and, because of this, HF and RFI were considerably reduced but not eliminated. Later developments in arc starting will be discussed in a following chapter.

A useful application of the Argonarc process at that time was for arc spot welding of thin sheet metals. It soon became obvious that DC TIG welding was highly suitable for joining most bright rolled steels, and particularly for many grades of stainless steel up to about 1.6 mm (1/16 in) thickness. A special spot welding gun was designed and introduced giving the facility to apply pressure and make welds from one side only where access was restricted.

Development of high efficiency resistance spot welding systems seems now largely to have phased out the TIG spot welding process although it is still to be found, often using the same hard-wearing original equipment which was being sold up to the mid-1960s.

It is probably true to say that the most extensive current use of TIG is in seam welding of stainless steels, where the DC mode is usually most suitable (precision TIG welding is almost exclusively DC). Aluminium, apart from a few suitable alloys, is still almost exclusively welded using AC, so many combined AC/DC sets have been developed. The efficiency of such units is well illustrated by the fact that many 25–30 year old TIG sets are still in use today, some of these not even having open-gap automatic arc strike but relying on the skill of an operator briefly to touch down and retract the electrode from the workpiece to initiate an arc before the electrode tip is deformed (touch start). Ever since its inception, equipment and consumables *e.g.* gas, wire, filler rods, *etc*, have been refined and improved although the basic principles of TIG welding have remained unchanged. A wide range of inert gas mixes has become available together with many varied alloyed tungsten electrodes. Power sources, mainly through advanced electronics, have become lighter, more accurate, and have greatly improved arc stability. This has enabled an increase in the use of TIG welding to be made, particularly for ultra-low current and precision applications. Control of weld sequences and programs by computer or microcomputer is now widely used and development in this field is still continuing. More details of power sources and their suitability for specific processes and materials are given later.

Chapter 1

The TIG welding process

Most forms of electric arc welding make use of the fact that shorting and then separating conductors connected to the positive and negative poles of an electric current source creates an arc, and thus an area of concentrated heat. If this arc occurs in air the metals being welded can become oxidised, and to some extent vaporised, by the uncontrolled intense heat produced. This unwanted effect can be reduced by use of a flux or, in the case of TIG, an inert gas shield. The power source thus controls the short circuit, directs the arc and allows the molten metal to flow evenly without oxidation.

Terminology

The process is known by various names as follows:

- *TIG* – Tungsten inert gas, the best known name in Europe but generally understood world-wide;
- *GTAW* – Gas tungsten arc welding, mostly in the USA;
- *WIG* – Wolfram (tungsten) inert gas, German definition.
  For the purpose of this book only the first definition is used.

Modes

- *DC* – direct current, electrode negative, work positive. Sometimes known as straight, *i.e.* unpulsed DCEN (or DCSP in the USA), Fig. 1.1.

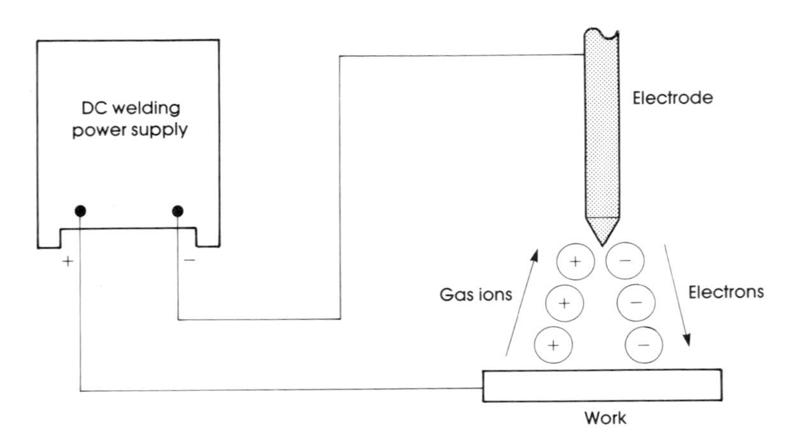

1.1 DCEN polarity.

- *DC reverse polarity*, electrode positive, work negative (not used as widely), DCEP (or DCRP in the USA), Fig. 1.2.

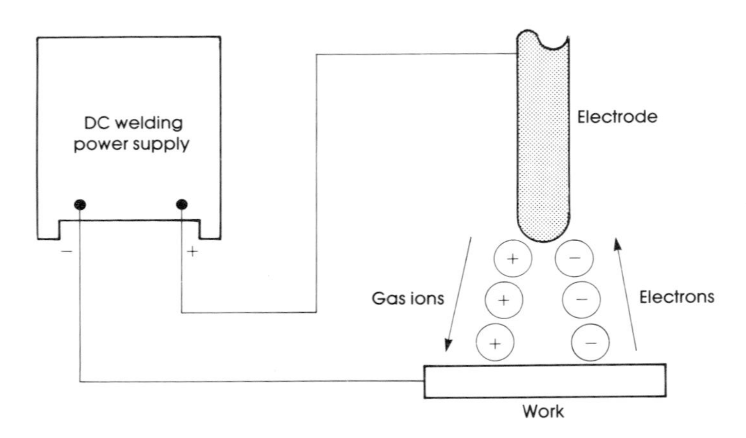

1.2 DCEP polarity.

Note: The terms DCEN and DCEP are preferable as they indicate electrode polarity for both the above.

- *AC* – alternating current. In this mode arc polarity rapidly changes giving some cathodic cleaning effect, ideal for aluminium where oxides rapidly develop in the weld bead, and for some stainless steels. An AC/DC power source gives the ability to select either DC or AC as the occasion arises and would be the best purchase for a general fabrication shop where many different metals of varying thickness to weld would often be encountered, Fig. 1.3.

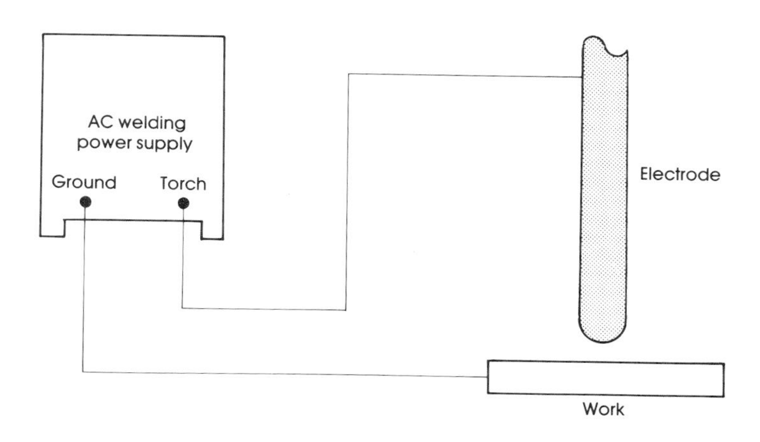

1.3 AC operation.

- *Pulsed DC* – This mode allows the arc to be pulsed at various rates between selectable high and low current settings and gives greater control of heat in the arc area.

TIG welding is clean, cost-effective, albeit a bit slow compared with metal inert gas (MIG) and metal active gas (MAG) welding, can be used by hand or automated and will weld a vast range of metals and thicknesses in several different modes. Whatever mode is used, the process remains the same, namely that the metals to be joined are fused together by the heat of an electric arc within a shield of inert gas which surrounds the arc and prevents undue oxidation of the metal. The arc is struck between the electrode and the workpiece and, in all but a very few cases, the electrode is made from tungsten, often with small quantities of a rare metal alloyed into the finished electrode rod.

A TIG arc is very hot and localised providing a means of applying maximum heat for welding in a small area, allowing an experienced welder to produce neat, compact weld beads with excellent penetration and strength.

Other terminology

AUTOGENOUS

A common term in TIG welding which means that the weld is formed by fusion of the parent metal(s) only and that no additional filler rod or wire has been introduced into the weld bead. It is generally considered that the *maximum* thickness for autogenous butt welding of mild and stainless steels by TIG is 2.5 mm (0.100 in) with other metals pro rata, depending on their heat conducting properties.

Note – a situation where the electrode touches the weld pool and welding ceases is known by expressions such as touch down, stub in or plough in, amongst others (often unrepeatable).

BACK PURGE

This is a condition in which additional shielding gas is piped to the underside of the weld bead, and some high quality power sources have an extra gas circuit and controls specifically for this purpose. It ensures that the underbead has a clean smooth surface with minimum or zero porosity. It is essential when fabricating vessels and welding tubing circuits for the food and drink processing industries where porosity could occur and harbour dangerous micro-organisms. It also serves to keep the underbead as small as possible, as large porous underbeads can affect the smooth flow of liquids through pipes and tubes.

CRATER

An imperfection or dimple at the end of a weld seam which can occur if the weld is suddenly terminated at full current. Craters can be eliminated, particularly in automatic welding by correct use of slope-down. With manual welding, craters are filled by use of the current control pedal.

DOWNSLOPE

Opposite of upslope, allowing the arc to die away or decay gradually. Also known as slope out and, in the USA, ramp out.

DUTY CYCLE OF A POWER SOURCE

A term used to define the period of time for which a power source can be used at a particular current level without overloading. Stated as a percentage, generally related to a ten minute period, or in the form of a graph, see Fig. 1.4.

A typical power source may be rated as follows:

Nominal rated* current 250 A at 60% duty cycle giving:
315A at 35% duty cycle . . . 3.5 min
* 250A at 60% duty cycle . . . 6.0 min
200A at 100% duty cycle . . . 10.0 min (continuous)

Welding shop use is nearly always intermittent, so the set usually has a chance to cool to normal operating temperature. Overload trips are often fitted to avoid electrical damage. These trips are either manually or automatically reset. When purchasing a set do not be misled by maximum current, pay attention to the stated duty cycle and select a unit with power above your normal needs.

Note: arc voltage generally increases slightly as duty cycle decreases. AC/DC TIG sets can be marginally less efficient than AC or DC *only* models.

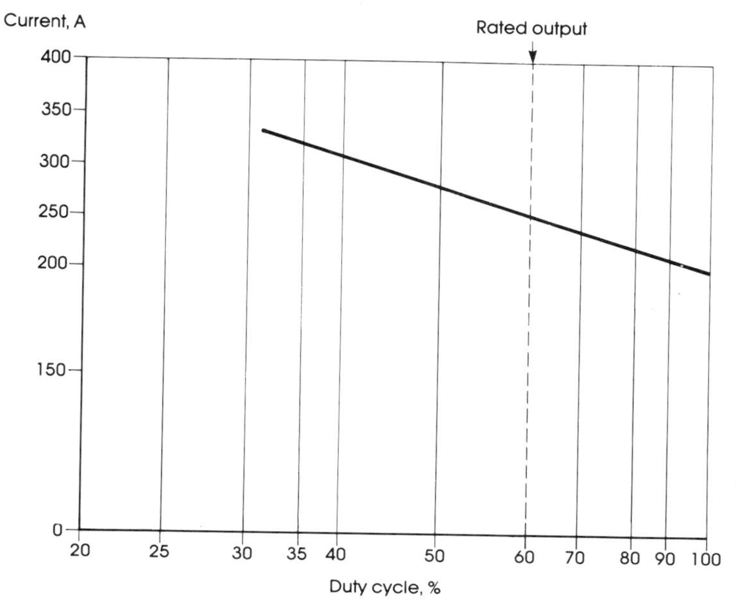

1.4 Duty cycle of a power source.

EARTH OR GROUND

These are misnomers but have become normal reference terms in all welding processes and should correctly be called welding 'current return'. The terms refer to the return circuit to the power source. Critical to the welding process, the return circuitry should carry the *full rated output* of the set.

HEAT AFFECTED ZONE (HAZ)

A term used to describe the region immediately around the weld bead. This area should be kept as small as possible and heat sinks are commonly used. Keeping the HAZ small reduces discoloration in the area around the weld zone and thus minimises finishing and polishing times.

HIGH FREQUENCY ARC START (HF)

In the past, HF pulses were used at the electrode tip to provide an ionised air bridge across which the welding current flowed from the tip to the work and established an arc. Once the arc was established the HF was terminated after a few milliseconds only. Now largely replaced by specialised electronic systems which eliminate, or at least minimise, RF and HF interference with peripheral equipment such as computers and DC motor control systems.

RF interference

RF stands for radio frequency and interference with electronics circuits and DC thyristor motor control systems can be caused by airborne RF. It generally results when electromagnetic waves combine either to reinforce each other or to cancel each other out, depending on their relative phases. It is rather difficult to eliminate when caused by HF arc start systems using spark gaps, but these are gradually being replaced by modern electronic devices which minimise both HF and RF interference. The effect of RF on welding system peripherals is similar to that caused by unsuppressed automobile ignition systems (very rare these days).

Computer programs have been known to be completely obliterated by both HF and RF but advanced technology has come to the rescue. Interference with TIG peripherals is usually caused by spikes from HF entering the electrical mains supply circuit.

NUGGET

A slang term, mostly used in the USA, to describe a complete weld bead, particularly between two adjoining, butt welded sheets.

OSCILLATION OF THE ARC

Sometimes known as weaving, this is lateral traversing of the arc from side to side across the seam over a short distance at a fixed speed whilst the torch travels longitudinally along the seam. It is best carried out by mechanical means, although highly skilled operators achieve excellent results by hand. Its purpose with heavier welds is to obtain good melt-in or sidewall fusion at the edges of the seam where the metal is thickest and it is particularly advantageous for capping runs when using a V or J edge preparation.

POST-PURGE

The condition in which shielding gas continues to flow through the ceramic nozzle on the torch for a period *after* the arc is extinguished. It is advisable always to allow a reasonable post-purge time as the electrode tip cools whilst still in an inert gas atmosphere, making for longer tip life between regrinds. The post-flow period also blankets the weld area whilst it is cooling, preventing undue oxidation and giving a better cosmetic appearance to the finished weld. The post-purge period is also often specified for certified welds.

PRE-PURGE

The condition is which the flow of shielding gas through the ceramic nozzle on the torch commences *before* an arc is struck. The period of pre-flow, usually variable at the gas valve on the power source, is mostly fairly short but serves to ensure that there is a minimum of free oxygen remaining in the arc area when welding commences. Extra long pre-purge periods are often specified for critical and certified welds.

UPSLOPE

Used extensively with thin metal sections, this is the condition where the arc is first struck at a low current and then increased on a timed basis up to the full required welding current level. It is ideal for automatic welding, particularly for critical circumferential or orbital applications. Also known as slope in and, in the USA, ramp up.

The following definitions do not strictly apply to this book but are included for interest:

GAS METAL ARC WELDING (GMAW)

The preferred term for MIG/MAG welding in the USA.

METAL INERT GAS (MIG)

Using argon or mixes for shielding.

METAL ACTIVE GAS (MAG)

Using CO_2 or mixes for shielding.

MOG

A rather ambiguous term. Stands for metal 0 (zero) gas although a shielding gas *is* produced from the heating of a flux core contained in the consumable wire electrode.

Note: the above are all semi-automatic processes using a continuously fed consumable wire electrode and a flow of shielding gas.

MANUAL METAL ARC (MMA)

The layman's general view of welding, using consumable stick electrodes with a flux coating.

SUBMERGED-ARC WELDING (SAW)

The weld is carried out using a wire fed through a gun similar to MIG/MAG but with the arc taking place under a continuously spread flux powder blanket (from a hopper) which can be recovered and reused to some extent. Useful for very heavy welds in thick carbon steels.

Chapter 2

Advantages and applications

From an operator's point of view the main advantage of TIG welding is the fine control of the arc. The process also produces very little spatter so that cleaning after welding is minimised. In addition, very little fume and smoke are produced during welding, although fume removal is recommended in most cases other than perhaps ultra-low current precision work. The TIG process lends itself readily to machine or automatic application, a subject which will be discussed in detail in a later chapter.

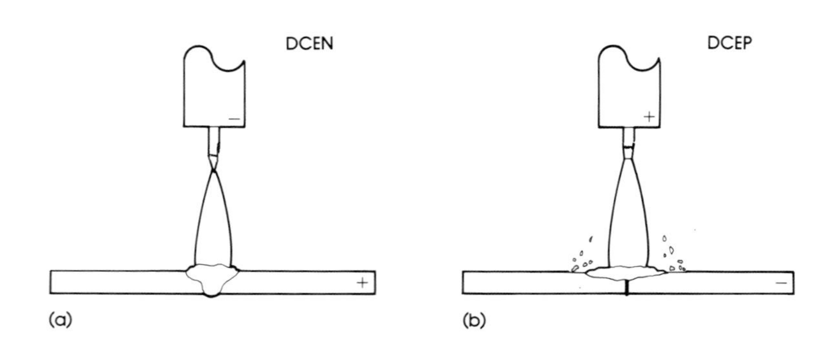

2.1 Arc effect: a) DCEN produces maximum workpiece heating; b) DCEP produces maximum workpiece cleaning.

Arc effect

When TIG welding the operator has three choices of welding current, two of which are shown in Fig. 2.1. They are direct current electrode negative (DCEN), direct current electrode positive (DCEP), and alternating current with high frequency stabilisation. Each of these current types has its applications and advantages and disadvantages. A look at each type and its uses will help the operator select the best current type for the job.

The type of current used has a great effect on the penetration pattern as well as bead configuration.

Typical applications for various metals

ALUMINIUM

Used in the AC mode the TIG process is ideal although some grades of aluminium alloys and pure aluminium are weldable using DCEN on thin sections. Heavy fabrications are possible, *e.g.* tanks, silos, pipes and structural framework.

STAINLESS STEELS

Heavy sections can be fabricated using the DCEN mode, giving a very good weld appearance which often needs no further fettling or polishing. Low current TIG is very suitable for the precision welding of stainless alloys in the manufacture of such items as bellows, diaphragms, load cells and transducers, often by mechanical means using pulsed TIG.

MILD STEELS

Not always as suitable for welding as the stainless alloys, mild steels are more prone to weld porosity and distortion.

Note: for both mild, carbon and stainless steels free-machining alloys can be a problem. Certain shielding gas mixes, special filler wires and rods and attention to heat sinking can often help with these metals but they should be avoided if high class welds are required.

Free-machining alloys can be taken as carbon and stainless steels with above average percentages of sulphur, manganese and phosphorus and lower than average carbon and silicon. These alloys *will* weld autogenously with careful attention to welding current pulse rate and heat sinking but are best avoided. They do save machining time and reduce tool point wear. However, check their weldability before embarking on batch production. The most common welding defects are brittleness and extensive porosity.

OTHER METALS: TITANIUM, ZIRCONIUM ALLOYS, NICKEL ALLOYS, COPPER AND ITS ALLOYS, TUNGSTEN

It may seem strange that tungsten can be welded using a tungsten electrode but it is, in fact, very weldable and is used in the fabrication of laser beam projector assemblies for example.

TITANIUM AND ZIRCONIUM

These are best welded and allowed to cool in a purge chamber to avoid oxidation and porosity, but they melt and flow easily to give a good cosmetic appearance. Both are considered to be reactive metals and great care should be taken to ensure good purging at all times. A trailing shield on the torch should be used in addition to the standard gas cup/nozzle assembly.

Some typical applications and components

- Bellows and diaphragm welding
- Transducer bodies and load cells
- Encapsulation of electronics
- Seam welds in tubes and sheets
- Sealing battery can tops
- Filter bodies and assemblies
- Jet engine blade and fin repairs
- Thermocouples and electronic probes
- Metal cabinet corners
- Valve housings and tube fittings
- Cladding and hard surfacing

Plasma versus TIG welding

It is not the intention of the writer to describe or discuss plasma-arc welding in this book. However, it could be useful to list a few of the advantages and disadvantages of plasma welding as compared to TIG.

ADVANTAGES OF PLASMA

- The tungsten electrode is sited well away from the work within the torch and is the cathode of an ionised inert gas column directed through a fine orifice in a copper nozzle. The plasma column thus formed is an incandescent, highly directional stream of gas. This means that coated metals can often be welded without removing the coating or unduly contaminating the tungsten.

- A pilot arc can be left running between bursts of full welding current.
- Metals above 3 mm thickness can be autogenously welded using a square butt presentation and the keyhole technique.
- Arc gaps are not too critical.
- The plasma arc is self-cleaning but careful cleaning of components should still be carried out.

DISADVANTAGES

- Plasma welding needs *two* different gas supplies. One, generally pure argon, is needed to form the plasma stream and the other, often an argon/hydrogen mix, is the shielding gas. This involves two sets of gas control equipment and very careful control of the gas flow rates.
- Watercooling of the torch is essential even at low currents to avoid overheating and erosion of the nozzle and orifice.
- Plasma equipment, particularly torches, is more expensive than that for TIG welding and needs more careful maintenance.
- A plasma arc is often too narrow and stiff, which does not compensate for component inaccuracies and leads to need for more accurate and thus more expensive fixturing. The slightly conical TIG arc can be a positive advantage in such cases.
- The plasma process is sometimes slower than TIG and often requires a second weld run, *e.g.* for bellows manufacture.
- Plasma equipment is generally bulkier and less portable.
- Nearly all bellows manufacturers have abandoned plasma and returned to TIG, particularly pulsed TIG.
- AVC (arc voltage control) cannot be successfully applied to plasma. For autogenous welds this does not always matter but it does if filler wire is being used. Surfacing build-up operations are much more easily carried out using TIG.
- Ninety five per cent of all microwelding can be just as satisfactorily carried out using TIG without the extra complications of plasma.
- Standard machine type torches are not readily available for plasma.
- Plasma cannot be easily applied to orbital pipe welding because of the additional complexity of pipework and cables to supply the weld torch with welding current and gas.
- Plasma arc starting nearly always needs HF which can interfere with computers, microprocessors and DC thyristor drives.

Note: Neither the TIG nor plasma processes are suitable for outdoor use except in very still air or with the welding arc securely curtained off to avoid draughts.

Summary

Whilst plasma scores on thicker and coated metals, nearly all microjoining operations are more easily and cheaply carried out using a precision pulsed TIG power source. For general welding TIG is more convenient.

For further information and advice on the plasma process see part 2 of 'TIG and plasma welding' by W Lucas, TWI, published by Abington Publishing UK, and other literature produced by The Welding Institute.

Chapter 3

Basic TIG welding requirements

The basic requirements for all TIG welding processes are similar, *i.e.* a power source, a hand held or machine manipulated torch, a pressurised supply of a suitable inert gas from cylinders or bulk containers and cables of the correct size to conduct welding current from the power source to the torch, Fig. 3.1.

This chapter covers torches, electrodes, shielding gases and cables.

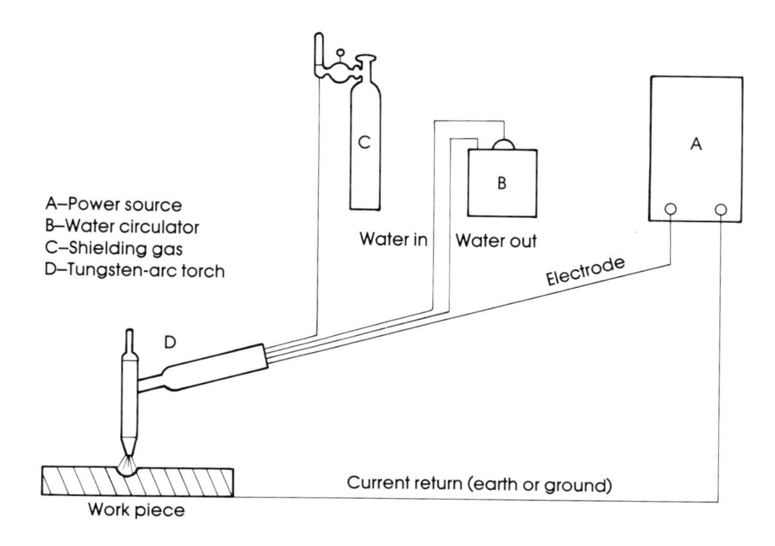

3.1 Basic TIG welding components.

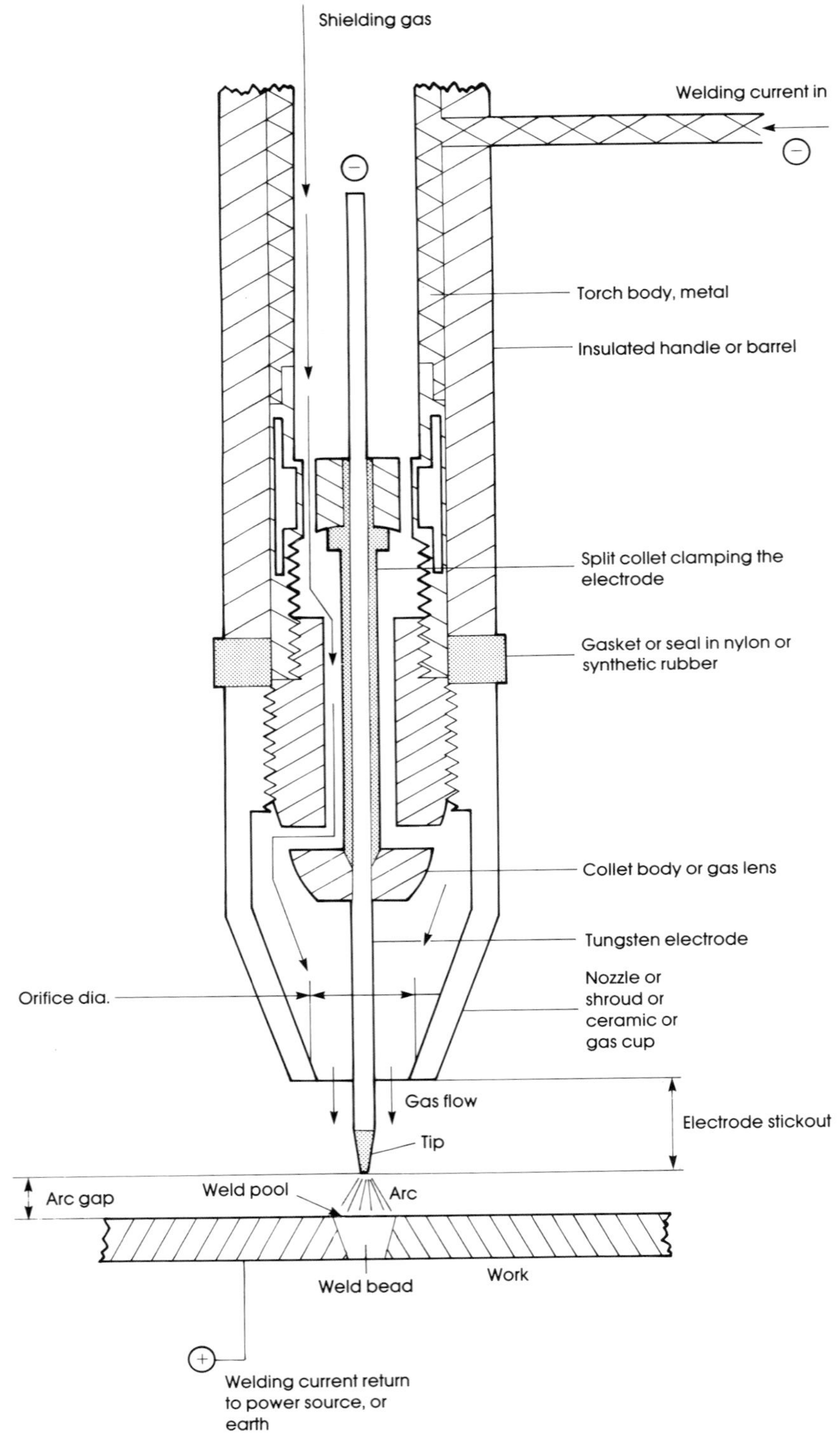

3.2 Gas cooled TIG welding torch assembly.

Torches

Delineation here is simple as there are only two basic kinds, hand and machine (straight). The torch carries the non-consumable tungsten electrode in an adjustable clamp arrangement and can be fitted with various ceramic nozzles to guide the gas flow to the arc area.

There has been little change in torch design since the inception of the commercial TIG process except that introduction of modern plastics and synthetic rubbers of lighter weight and with superior insulating properties has allowed torches to be made smaller, lighter and easier to handle. Manual TIG welding requires a steady hand, so a lightweight torch is a great advantage. However, torches should always be selected with due regard to capacity. Use a torch which will carry the *maximum* welding current that is likely to be needed, a manufacturer or stockist will give assistance, and ensure that the nozzle and electrode assembly allow full access to the area being welded. It is, of course, preferable to weld in the flat (downhand) position to take full advantage of the effect of gravity. This is particularly so when additional metal is being introduced into the weld pool by means of wire or rod. Further advice on suitable approach and position will be given later.

The various components of TIG torches are shown in Fig. 3.2, 3.3 and 3.4 which also give the most used names of the respective parts.

An extremely large variety of ceramic gas cups is available and it is often difficult for beginners to decide which is best for a particular job. Ceramics are graded by orifice size and a rough guiding rule would be that the smaller the electrode and the lower the gas flow the smaller the orifice. Gas flow rate depends on the type of gas and cup orifice diameter. For most hand welding about 7 l/min (15 ft^3/hr) is a good amount.

Always keep a selection of nozzles handy with a good range of orifice sizes, Fig. 3.5. The flow of shielding gas through the torch has some cooling effect which is sufficient for low production rates. For reasons of economy, however, it is better that the gas flow is kept to a minimum, but it reduces the cooling effect. This means that the torch could become uncomfortably warm during continuous use and need watercooling. Torches above 125 A capacity are generally available as either gas- or watercooled. For a busy fabrication shop watercooling can be essential.

Figure 3.2 shows the component parts of a TIG torch and also indicates the usual position of the watercooling gallery. In some cases the water is directed in and out of the torch by separate tubes, but it is not uncommon for the cooling water to be concentrically piped away from the torch via the current input cable, which cools the cable as well. The flow of cooling water needs only to be fairly small (about 1.5 l/min) and for reasons of economy it is best if the watercoolers are of the recirculating type. Mention of these cooling units will also be made in a following chapter.

Manufacturers have taken much trouble to ensure that TIG torches are

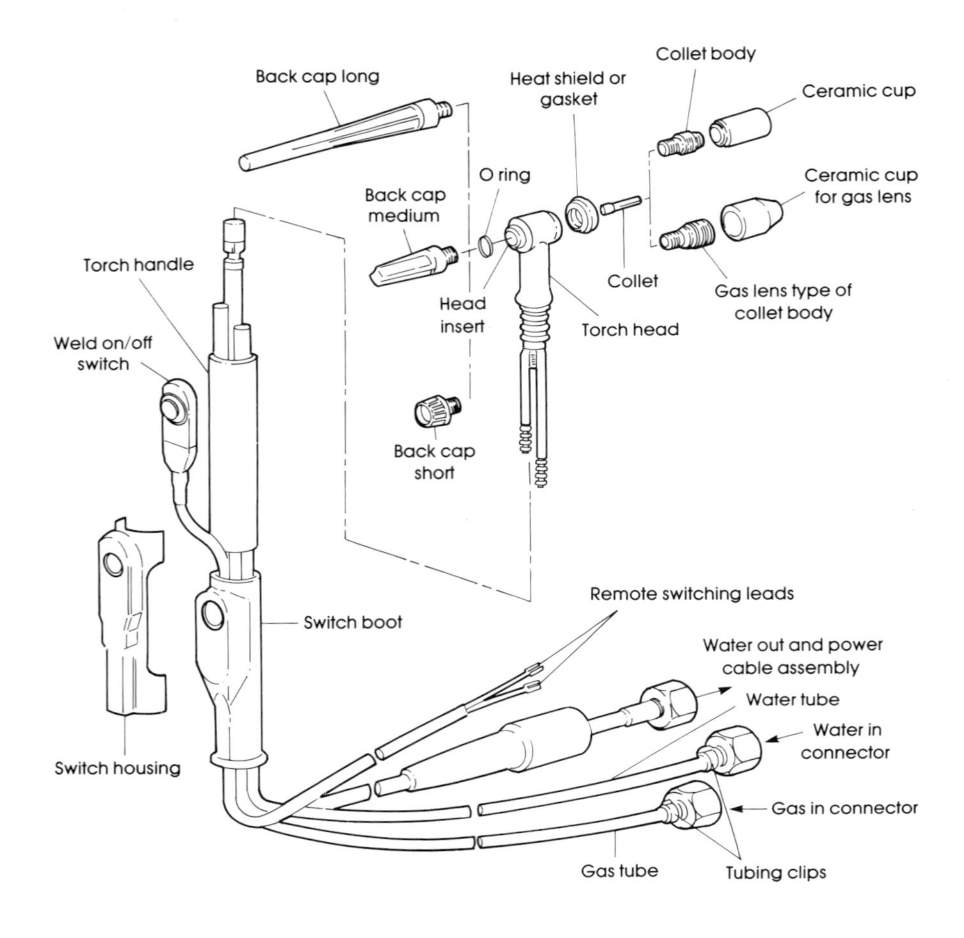

3.3 Watercooled TIG welding torch and fittings.

easy to handle and also that the various component parts are easily changed or replaced. On most designs the ceramic nozzles can be changed in a matter of seconds and it should not take much longer to change or replace the tungsten electrode.

The metal component parts of a good quality TIG torch should always be made from copper or a good quality low resistance brass.

To summarise, when choosing a TIG torch always ask yourself:

- Does the torch have sufficient capacity for the maximum welding current to be used?
- Is it comfortable for the operator if hand welding?
- Is it suitable for welding where access is awkward or restricted?
- Is the cooling method adequate for the production rate required?

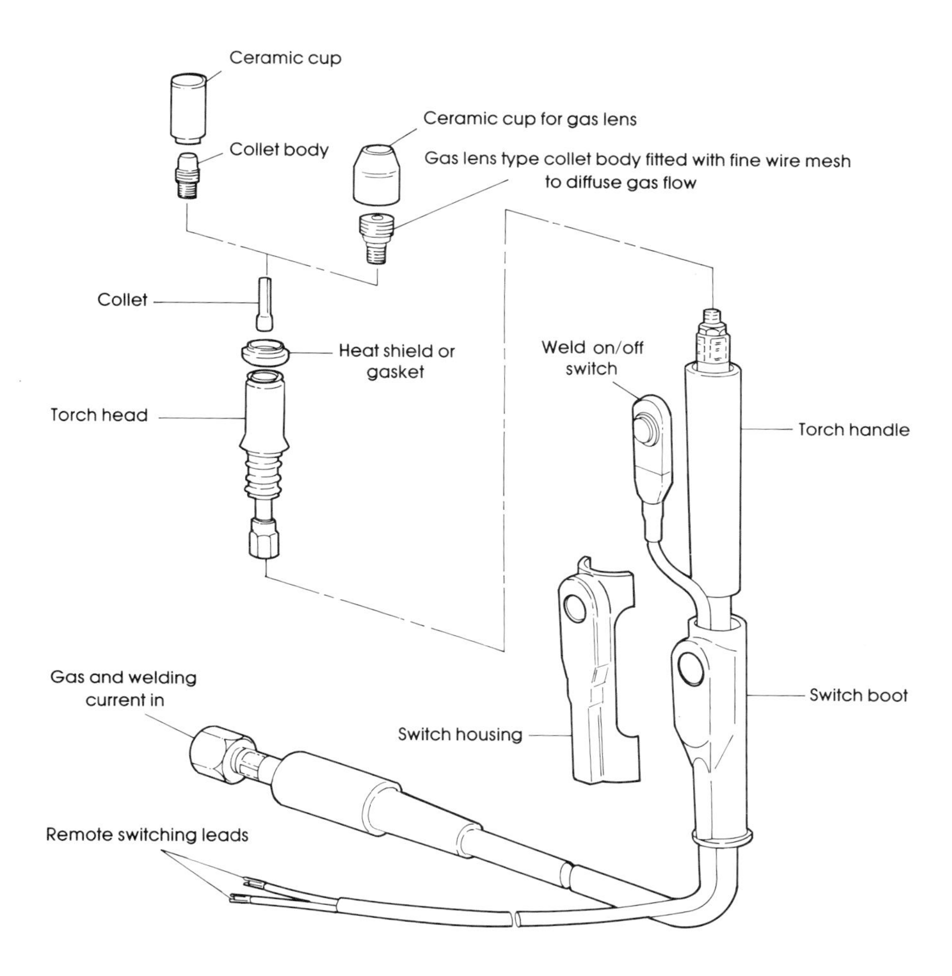

3.4 Gas cooled TIG welding torch and fittings.

- Does the nozzle-to-body sealing method prevent air ingress into the gas flow? At this point a synthetic rubber seal is better than nylon or similar material.

The flow of current to a TIG torch used in hand welding is usually controlled by a pedal switch, leaving one hand free to apply filler wire if necessary, or by a small push button on the torch itself which only switches the weld current on and off.

There are also some TIG torches available which have a flexible neck. These are useful where access to the weld is very difficult, although they are not otherwise extensively used.

TIG torches are fairly robust but should be kept clean and free from oil in particular. Lay the torch down carefully when not in use or, better still, have

3.5 a) Selection of manual TIG welding torches and gas cups; b) TIG torch with built in wire feed and operator's stop-start button.

an *insulated* hook handy to hang it on. This will keep the torch ready to hand and prevent it from falling to the floor, a simple tip which could save the cost of a replacement unit. Keep all torch connections, *i.e.* welding current, shielding gas and cooling water, tightened. A loose current connection within the torch can cause internal arcing with subsequent overheating and, obviously, gas and water leaks are *not* wanted. With care a torch can last a long time and give good service throughout its working life.

Electrodes

TIG electrodes are the final link in the chain between the power source and the weld and critical to the whole process. They are classified as non-consumable, which is not strictly true. What *is* true, however, is that they should last as long as possible for economic reasons, that they should carry the highest required welding current and that they should not disintegrate at the tip whilst welding, thus contaminating the weld pool. All this makes the selection of a suitable electrode of paramount importance. Electrodes for TIG welding are almost entirely made from tungsten (which has a melting temperature of 3370 C (6000 F) and boils at 6135 C (11000 F). They are generally obtainable in round rod form about 150 mm (6 in) long. The rods are mainly products of powder metallurgy produced by compression and/or sintering and come in a variety of diameters starting at about 0.25 mm (0.010 in) up to 6 mm (0.25 in). Larger diameters can be obtained if required. Table 3.1 gives some advice as to which size to use for a particular weld current range and is based on 2% thoriated electrodes, although ceriated and lanthanated tungstens have very similar characteristics. Table 3.2 gives electrode weights for specific diameters.

Table 3.1 TIG electrode current carrying capacity. Tungsten 2% thoriated, 60% duty cycle, in argon

Diameter, in		Diameter, mm	DCEN or DCSP	DCEP or AC	Popular sizes
			Current capacity, A	Current capacity, A	
	0.010	0.25	up to 5	N/A	
	0.020	0.5	up to 15	N/A	✓
3/64	0.040	1.0	15–50	up to 20	✓
1/16	0.064	1.6	50–100	20–50	✓
3/32	0.080	2.0	50–150	50–100	✓
	0.096	2.4	50–200	50–150	
1/8	0.128	3.2	200–300	150–200	✓
3/16	0.180	4.8	250–400	200–300	✓
1/4	0.250	6.4	400–600	300–400	✓

For greater currents consult your electrode supplier

Table 3.2 Nominal weight of standard length tungsten electrodes

Standard diameters, mm	Nominal weight per electrode in grams for standard lengths of 75 mm (3 in)	150 mm (6 in)	175 mm (9 in)
1.0	1.1	2.3	2.7
1.2	1.6	3.3	3.8
1.5	2.6	5.1	6.0
1.6	2.4	5.8	7.0
2.0	4.5	9.1	10.6
2.4	6.5	13.1	15.3
3.0	10.2	20.5	23.9
3.2	11.7	23.3	27.2
4.0	18.2	36.4	42
4.8	26.2	52	61
5.0	28.4	57	66
6.0	41	81	95
6.4	47	93	109
7.0	56	111	130
8.0	73	145	170
10.0	114	227	256

To improve the flow of electrons through the formed rod to the tip it is common practice to include in the tungsten electrode powder mix small quantities, seldom more than 4%, of various metallic oxide powders mixed and distributed as homogenously as possible within the structure of the rod. These include oxide powders of zirconium, thorium and, latterly, lanthanum and cerium in various percentages. All these inclusions greatly improve arc striking, particularly when low current DC welding is being carried out and assist with tip shape retention and arc stability. Thorium dioxide (ThO_2) has until recently been the most common oxide used and for most purposes is still by far the best. However, factors other than long tip life and thermal efficiency now weigh against its use and equally good substitutes have been made available. In order of preference for arc striking only, in the author's opinion and based on several years' usage and experiments, are:

1st Thorium 4%, 2% is also satisfactory;
2nd Cerium 1% (cerium is the most abundant lanthanide element);
3rd Lanthanum 1% (lanthanum is the first of the series of rare earth metals).

With regard to lasting properties during DC welding there is not much difference between these three but experience shows that thoriated electrodes last longer between regrinds for currents above 100 A, with all the three lasting well at currents below this figure. For ultra-low current welding (below 3 A) the author's personal preference is for the ceriated type which strikes well and holds a very fine point for long periods. Zirconiated (0.8%) electrodes are mainly used for AC welding as are those of pure tungsten.

IDENTIFICATION OF ELECTRODE TYPE

Electrodes are identified by colour and a DIN standard has assigned a colour code. However some of the colours, particularly orange and the pinks and reds (marked*) are often difficult to distinguish one from the other, so be very careful when a particular type *must* be used. Table 3.3 shows the colours suggested in the DIN standard which, however, is not as yet universally accepted.

The attempt to colour code electrodes by a standard is an excellent idea but for this to become universally used some acceptable international standard should also be assigned to the various colours to avoid confusion.

Table 3.3 Electrode data to DIN 32528 specification

General use	Composition, %	DIN	Material No.	Colour
AC	Tungsten, pure	W	2.6005	Green
AC and DC	Tungsten +1 thorium	WT10	2.6022	Yellow
DC	Tungsten +2 thorium	WT20	2.6026	Red*
DC	Tungsten +3 thorium	WT30	2.6030	Lilac
DC	Tungsten +4 thorium	WT40	2.6036	Orange*
AC	Tungsten +0.8 zirconium	W28	2.6062	White
DC	Tungsten +1.0 lanthanum	WL10	2.6010	Black
DC	Tungsten +1.0 cerium	WC10	—	Pink*
DC	Tungsten +2.0 cerium	WC20	—	Grey

* *Colours not easy to distinguish*

WEAR AND LIFE PROPERTIES

The two histograms, Fig. 3.6, were produced from results of experiments carried out by a major manufacturer of tungsten TIG electrodes. Four different types of alloyed rod were used, all 3.2 mm (0.125 in) diameter at a continuous current of 150 A in both DC and AC modes. The results appear to confirm the author's opinions with regard to tungsten life, and the fact that ceriated electrodes last well in relation to the thoriated type removes many problems regarding dangerous grinding dust and its disposal.

Assuming that the results were obtained with a fixed electrode to work distance, or arc gap, a consistent flow rate and identical TIG torches, the data would probably apply only to mechanised welding. Hand welding in a busy fabrication shop can give very different results.

ARC GAP

A TIG arc is extremely hot, about 20 000 C (35 000 F). This hot spot occurs in the arc at a point close to the electrode tip, but not in the electrode itself, so for optimum penetration and economy the arc gap must be kept as small as is practicable, fairly easy to maintain in mechanised TIG welding but entirely due to the steady hand of the operator in manual TIG. The width of the arc gap has considerable effect on the amount of heat going in to the

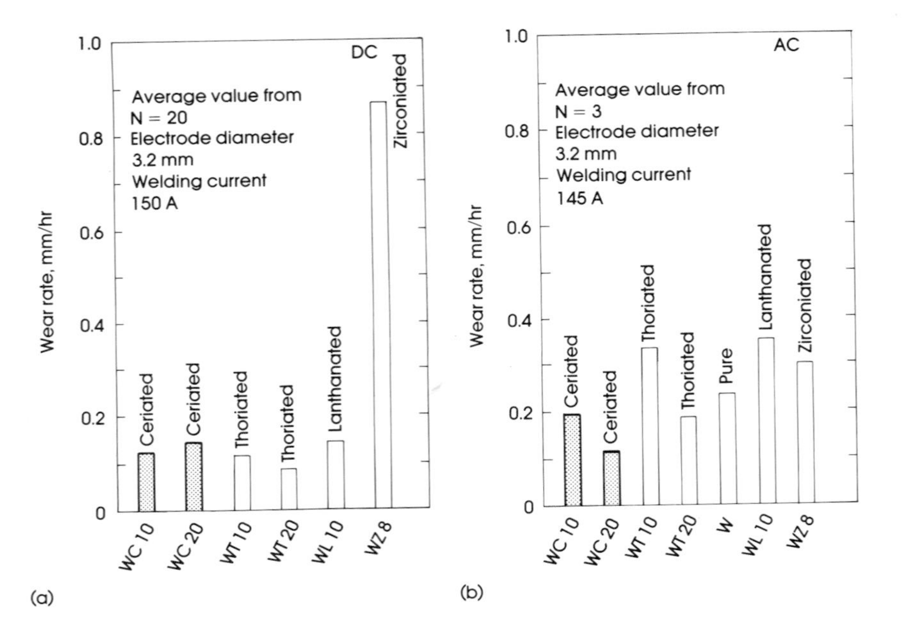

3.6 Wear rate for various electrode compositions when TIG welding: a) Steel with DCEP polarity; b) AlZnMgl with AC.

weld pool, as variations in the gap give variations in total heat. The relationship is simple:

Amps × volts × time (seconds) = joules (or watt-seconds).

Where filler wire is being used the arc gap must be greater than for an autogenous weld. Also, the wire or rod has to be melted, so a weld *with* filler will need at least 20% more current than one without. The heat input for any TIG weld can easily be found by experiment but experience will often indicate what approximate current is required before welding starts. No attempt has been made in this book to advise on arc gaps, keep them as small as possible.

The passage of current through the electrode and the heat of the arc will eventually cause the tip to degrade. Some metals out-gas during melting and this can occasionally coat the tip, making repeat arc strikes difficult to achieve. In fact some of the more exotic metal alloys can cause coating of the tip to such an extent that a state of one weld, one regrind exists. When the electrode tip has degraded or become coated it is time to regrind.

ELECTRODE POINTS – TIP GRINDING

Generally the rule here is: the higher the weld current the larger the electrode diameter and the greater the included tip angle, within a 30 – 120° range for most TIG welding use. Tips with an included point angle as low as 10° or less are used for such precision TIG applications as the fusion of thermocouple wires and for edge welded bellows and diaphragms. Regarding grinding of electrode points, the surface finish of the tip is critical only in precision and microwelding applications and in these instances it is a definite advantage to use a mechanical tip grinder. These machines give finely finished, and sometimes polished, tips with first class repeatability and are simple to use, although a good one can be rather expensive. Tungsten grinders will be further discussed later. A capable TIG welding operator rapidly learns to grind tips to an acceptable standard for hand welding on a standard bench grinder but this *must* be fitted with suitable hard, fine grit wheels and an eye shield. Tungsten is very hard and grinding will rapidly destroy softer wheels with subsequent dust problems and increased replacement expense. Purchase of grinding wheels suitable for prolonged use with tungsten electrodes will save money in the long run; consult a reputable supplier.

One final comment. To assist the welding current electrons to be emitted from the very tip of the electrode point, ensure that any marks left by grinding run longitudinally *towards* the tip, *not* circumferentially, (Fig. 3.7) and keep points concentric with the tungsten shaft centreline.

GRINDING DUST

Dust generated by grinding can be considered a severe health hazard and most countries' health authorities recommend that all metal grinding machines be fitted with a transparent eye shield, filter and dust extractor. Where tungsten grinding is concerned, thorium, although present in only small quantities, has two particularly unwanted characteristics, *i.e.* it is slightly radioactive and is also regarded by some medical authorities to be carcinogenic. All dust created by the grinding of thoriated electrodes should be collected and disposed of with extreme care. In the UK an addition of thorium greater than 2% brings these electrodes into the province of the 1985 radiation regulations. *No* dust produced by grinding should be breathed in, as any dust can cause pulmonary illness over long periods of inhalation, so it is only commonsense to provide greater protection. Always wear goggles or industrial spectacles when grinding anything. If there are any doubts about the efficiency of a grinder's dust extraction system, operators should also wear face masks.

Shielding gas types and mixtures

A wide variety of inert gases and mixes of gases is available, either in pressurised cylinders or as liquids contained in special insulated bulk tanks. If large quantities of gas are being used the bulk tank is the most economical method but, of course, such tanks are for static mounting and are not portable.

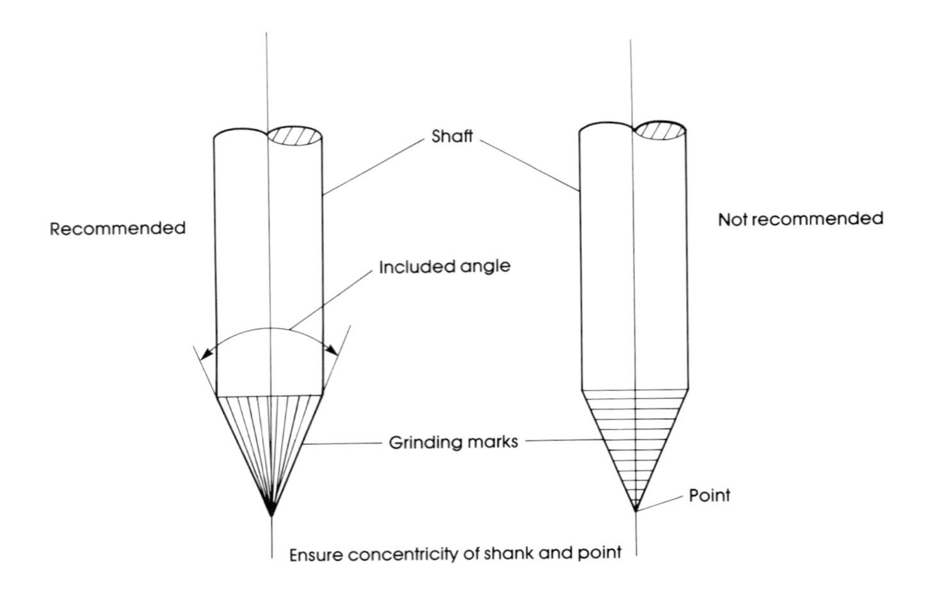

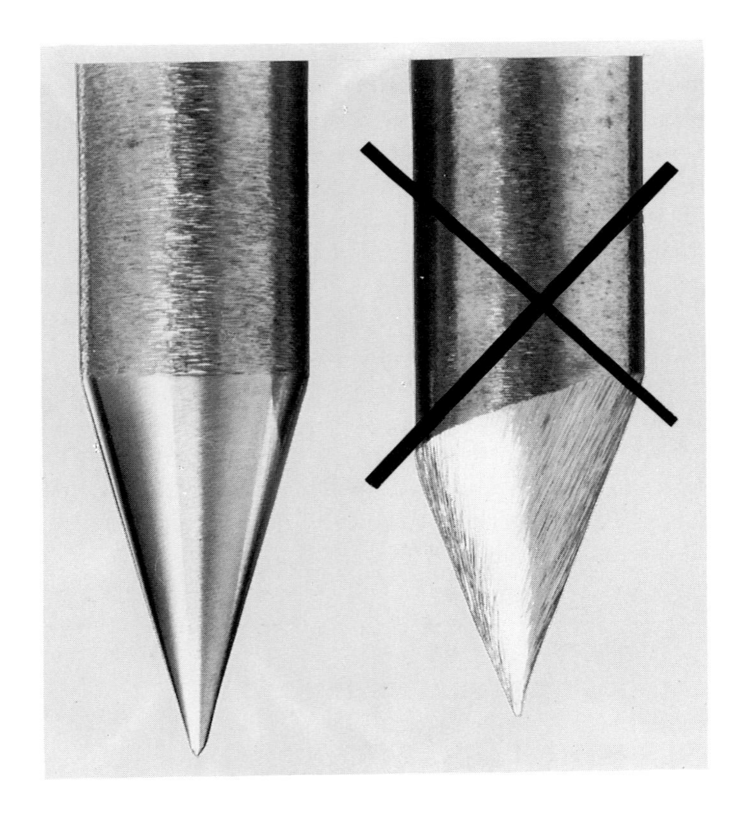

3.7 Correctly and incorrectly ground electrode points for DC TIG welding.

These liquid gases are held at very low temperatures, in the case of argon at −300 F, so the siting of bulk cryogenic tanks is critical. Consult your supplier for advice on which is most suitable for you. Notes on the various gases now follow:

ARGON

This is the most common and economical gas used in TIG welding and is obtained as a by-product when air is liquefied to produce oxygen; argon being present in air at about 0.9%. The advantages of argon are its low ionisation potential and thermal conductivity. Also, being about 1.5 times denser than air it maintains good blanket coverage at the arc for longer periods than helium.

The ionisation potential of argon is around 15.5 electron volts. This is the voltage necessary to remove an electron from a stable atom and convert it into a charged atom or positive ion, making the gas in the area of the arc into a plasma. The remainder of the shielding gas around the arc excludes the active components of the surrounding air and prevents, or at least minimises, metal oxidation.

Argon gives a high arc energy density, *i.e.* a high concentration of energy within the arc area, Fig. 3.8. This allows production of narrow weld seams and it can be obtained at a purity of better than 99.90%, which is essential for welding. It can be used for a great variety of metals and is particularly good for mild and stainless steels, but equally useful for aluminium and magnesium alloys.

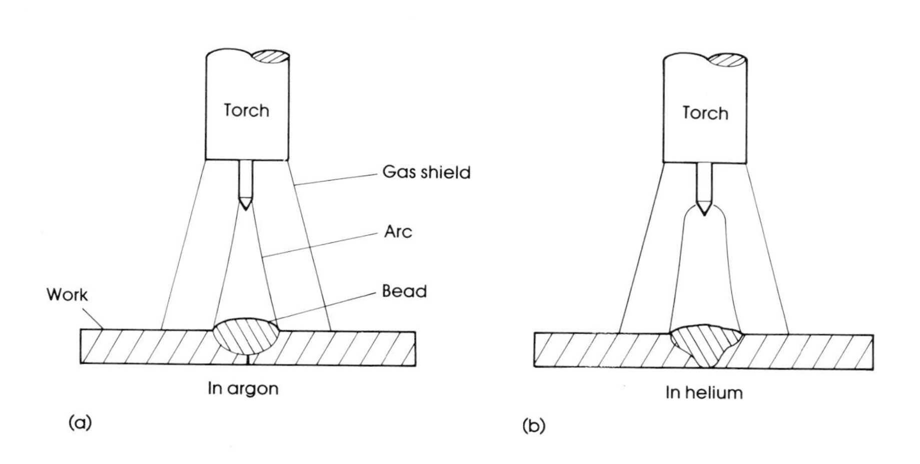

3.8 Arc shape using: a) Argon; b) Helium.

A range of argon mixes is available combined with hydrogen additions of between 1 – 5%. These gases concentrate the arc and can increase welding speed but they are more often used to give a better appearance to the finished weld. Hydrogen is classed as a reducing gas. It should be noted that inclusion of hydrogen can give rise to weld porosity, so the rule here is to use the lowest ratio of hydrogen to argon consistent with a good weld both for strength and appearance, Table 3.4. Your gas supplier will assist with the choice of mix.

Table 3.4 Shielding gases for various metals

Gas mix, %	Mild steels	Low alloy steels	Stainless steels	Nickel alloys	Aluminium and alloys	Copper and alloys	Remarks
Commercial argon 99.995	•	•			•	•	General use
High purity argon 99.998	•	•	•	•	•	•	Fine precision welding
Ar75/He25	•	•	•		•	•	Very suitable for aluminium
Ar70/He30	•	•	•	•	•	•	Very suitable for aluminium
Ar50/He50	•	•	•	•	•	•	Very suitable for aluminium
Ar99/$H_2$1	•		•				Not for use with martensitic s/steels
Ar98.5/$H_2$1.5	•		•	•			Not for use with martensitic s/steels
Ar98/$H_2$2			•	•			Not for use with martensitic s/steels
Ar97/$H_2$3			•	•			Not for use with martensitic s/steels
Ar95/$H_2$5			•	•			Not for use with martensitic s/steels
Special Ar/H_2 mixes			•	•			Not for use with martensitic s/steels
Commercial He 99.993				•	•	•	Not for use with any steel
Ar/N						•	Not for use with any steel

HELIUM

Helium is an ideal shielding gas but more expensive in the UK and Europe and not therefore so widely used, in particular for hand welding. Its ionisation potential is 24.5 electron volts, with excellent thermal conductivity, and it gives deeper penetration than argon for a given current and arc gap.

Figure 3.8 shows the general TIG arc forms obtained with argon and helium when used for *mechanised* welding.

With an equivalent arc length, helium has a higher arc voltage than argon. The relationship (amps × volts × time = joules) shows that an increase in arc voltage considerably increases heat input to the weld, so helium or mixtures of argon and helium will score when welding thick metals or using high travel speeds, whilst minimising porosity. However, arc lengths must be held to a very close tolerance when using helium and this means that it is more suitable for machine welding, seldom being used for hand welding unless by a highly skilled operative.

Other disadvantages of helium are the necessity for high flow rates, with an accompanying increase in cost, and the fact that it can inhibit arc striking at low currents. Above 150 A it has some distinct advantages for thicker metals.

Helium/argon mixes are available with argon percentages ranging from 50–75%. These are ideal for most metals and alloys, particularly copper and aluminium.

When used in an automatic welding system helium can, in some circumstances, entirely inhibit an open-gap arc strike and will almost certainly have an adverse effect on arc pulsing. If it is absolutely necessary to use pure helium, a solution could be first to establish an arc in high purity argon and then rapidly change to helium when an arc is established.

NOTES:

- The purity percentages are for gas in cylinders only. Bulk and liquid gases tend to be less pure. Consult your supplier;
- Gas mixes containing hydrogen require a special cylinder regulator usually with a left hand (LH) connecting thread. Again consult your supplier;
- Keep cylinders at room temperature (about 20 C) when welding. If your cylinder storage area is colder than this, allow cylinders to warm to room temperature *before* using;
- Ensure that all piping and connectors are airtight and moisture free. Argon in particular has a slight tendency to be hygroscopic and can collect slight traces of water vapour through piping, leaks, *etc*;
- Argon with 3 or 5% hydrogen is particularly suitable for automatic use, giving increased heat input and thus higher welding speeds;
- Argon with 1% hydrogen gives a clean and shiny cosmetic appearance to the weld bead or seam, a big advantage to the appearance of instruments, *etc*. Major gas producers and suppliers produce Ar/H_2 mixes at a customer's request to his specification. Gas producers also give advice on gas mixes for specific applications;
- Argon with a small addition of nitrogen can help in welding copper as it increases heat input at the arc. *Never* use an argon/nitrogen mix with

ferritic metals as it reduces weld quality and strength. If in doubt consult your gas supplier;

- For very special purposes a gas supplier can provide a mix of more than two gases but this is unusual and costly for small amounts.

BACK PURGING

The finish of the underside of a weld bead is as important as the top surface and can be improved by back purging using the same gas as that flowing through the torch, see later chapters.

Cables

In general the only independent cable used in TIG welding is the current return (earth) and by far the best for this purpose is stranded copper, covered with a natural or synthetic rubber (rather than plastic) which will remain flexible over a wide temperature range. Use good quality cable from a rep-utable manufacturer and keep to the maximum current capacity stated. Always ensure that the connectors at both ends are tight and will not work loose in use. Keep cables as clean as possible and away from possible damage by welders' boots, fork lift trucks or heavy components.

TIG welding produces very little spatter and a suitable natural or synthetic rubber covering is the most resistant to heat damage. PVC and other plastic coverings have adequate insulation properties but are more prone to heat damage and are not so flexible, particularly in cold conditions.

COMPOSITE TIG TORCH CABLES

Composite current cables are often the most troublesome part of a TIG welding system often through failure of the connections, allowing water leaks or the introduction of air into the shielding gas. Many users buy torches with regard only to cost which usually dictates the type of cable fitted. For gas cooled torches there are two styles of cable available, the first a two piece type which uses insulated welding cable to carry the current and a separate hose for the shielding gas. The main advantage with this system is that it is easy to repair and generally long lived. To its detriment are loss of flexibility and comparatively high cost. The most popular option is the use of monocable incorporating the copper conductor inside a larger hose through which also passes the shielding gas. This has the advantage that the hose acts as an insulator for the copper strands while simultaneously supplying shielding gas, and flexibility is improved considerably although ease of maintenance is lost. Monocables are sold in three main forms: PVC hose, reinforced PVC hose and overbraided rubber hose. Rubber hose is by far the superior material combining excellent flexibility with unsurpassed heat

Table 3.5 Sheaths and coverings for composite cables

Material	Rating	Comments
Natural rubber sheathing	★★★	Good all round performance; resists abrasion well and protects against spatter and grinding sparks. Swells in contact with oil. Thicknesses less than 1.5 mm tend to tear in use.
Neoprene sheathing	★★★★	Has similar properties to natural rubber. Less liable to tear and is self-extinguishing.
Unreinforced PVC monocable	★	Low quality: very likely to leak when hot. Flexibility poor, especially when cold.
Reinforced PVC monocable	★★	Reasonable resistance to leakage when correctly crimped. Good flexibility. Be wary of operating gas cooled torches at maximum current for long periods.
Two piece power cables (rubber or PVC)	★★★	Easily maintained. Not as flexible as monocables but hardwearing especially when rubber coverings are used.
Rubber monocable (natural and synthetic)	★★★★★	Extremely flexible, hardwearing and unlikely to leak. *Note*: rubber hoses should always be overbraided or reinforced as plain hose might burst when pressurised.
Nylon zip covers	★★★★	Very flexible with good abrasion resistance. Protection against oil, water and spatter not good.
Glass fibre zip covers	★★★	Very good heat resistant and flexibility, poor life in abrasive conditions.

resisting properties. The PVC hose suffers from heat ageing which in certain conditions rapidly causes leakages around the crimped areas, especially with unreinforced hoses, Table 3.5. The choice, which will probably depend upon the finance available, should also pay attention to the cable ends to ensure that the outer ferrules are of a length to give an effective and reliable water/gas seal. This is the most common point of failure. A 20 mm long ferrule can be considered to be sufficient.

With watercooled torches the construction of current cables is usually of the mono style. In these the cable acts as the drainpipe for the cooling water. The effect of this is twofold; heat ageing of the hose is not such a problem and the cross sectional area of the copper strands in the cable can be reduced because of the watercooling. The weight reduction is also an advantage when using watercooled torches in comparison with similarly rated gas cooled torches which require a much heavier cable. Again the use of rubber hose is of benefit to flexibility and torch life.

SHEATHS

Although it adds weight an overall sheath should be considered for composite power cables to provide protection for the leads and prevent damage as they are dragged across workshop floors. Sheaths come in many forms, dependent upon the torch manufacturer, the best type being made from rubber, fitted at the point of manufacture. This gives good resistance against abrasion and spatter, dependent upon the material. It may also be fire retardant. Available as a retrofit item is a zipper cover which is normally made from a nylon or glassfibre material and is zipped up over the length of the cables. This type is generally more expensive but has its advantages in use as the cover can easily be removed and replaced.

The choice of cables to be fitted to a TIG torch is dictated first by the application and second by price. Table 3.5 gives short lists of available types as a rough initial guide to price and performance.

The star rating indicates the merits and wearing properties of various materials for cable coverings, the more stars the better.

CABLE CONNECTORS

Table 3.6 gives brief descriptions of type and use.

Most countries' standards associations have recommendations both for cables and connectors but considerable confusion still exists as to the production of a *world* standard for such items. The associations meet regularly to try and resolve the matter (particularly European and American) but at the time of writing no firm agreement has been reached. Until such a happy day arrives, use suitable conversation units and adapt as required.

Table 3.6 Cable connections

Type	Use	Description	Comments
BSP female	Mainly European power sources	Standard end fittings of a basic type as stocked by most distributors	Tighten well and insulate with a suitable rubber or plastic boot
American fittings	USA imports to Europe	Torch leads can be made up to this specification for the UK but often at a high price	Normal method used in the USA and readily available there. High current capacity
Central adaptor	Fitted to a number of European machines (Advantage: quick torch connection)	Produced by a number of manufacturers but pin size and position may vary	Quick fit and release. ESAB proprietary type, *etc*
Central connector	Most European power sources and cable extension connectors	Available to order. Ensure that the correct type is specified. A number of different types are available rated on size and current capacity	Quick fit and release. Becoming increasingly popular in Europe. BINZEL proprietary type plus DIN, *etc*
Stud fittings	Older UK and USA machines	Use standard torch cable adaptors (spade type)	Tighten well and insulate
Dual purpose	UK and USA machines easily available	Carry both welding current and cooling water	Very popular for USA machines, orbital weld heads, *etc*

Whatever type of fittings are used, great care must be taken to ensure that *all* cable connections are tight and well insulated with no bare live metal parts. In particular check that your connectors are insulated against HF leakage when HF is being used, by ensuring that rubber boots, *etc*, are adequate, fit tightly and overlap all mating parts. When clamping fittings to cables all wire strands must be clamped by the ferrule. A good ferrule is essential to eliminate stray arcing and overheating at the clamped joints.

CABLE MARKING

Good quality cable should have its size and current rating printed on the insulated sheathing at intervals along its length. However, many equally good cables do *not* have this marking so always buy from a reputable supplier and hold him responsible if not satisfied.

Chapter 4

TIG welding power sources

The power source is the heart of all welding systems; reliability, accuracy and long life being the desirable characteristics governing the selection of a set. Price does, of course, enter into the choice but with TIG power sources, as with any electromechanical device, you get what you pay for. When considering purchase, pay great attention to duty cycle, spares availability and the speed with which servicing can be obtained if required. In small workshops there is seldom a back-up set, so choose wisely. An enormous range is available and most reputable suppliers will arrange a free trial.

The power source must convert mains electricity, with its inherent variations, into as stable a welding current as possible. It does this by using a transformer to reduce mains voltage and proportionally increase current through the secondary windings and convert this to welding current using a rectifier. Early welding sets were bulky, heavy and often unreliable, although it is remarkable how many old sets are still giving good service, probably because of their heavy, sound, construction and relatively low duty cycle. Duty cycle is measured on the basis of how long maximum current can be used over a percentage of a given period, without overheating or internal damage. Modern sets now use overload trips, *etc*, to minimise damage.

Modern power sources are much lighter and less bulky through use of solid state electronics, efficient cooling and modern materials such as plastics and light alloys in their construction. It is generally considered that the four main basic types are as follows:

Type 1 Transistor series regulator power sources - DC only

These use power transistors for current regulation, with analogue control from a low current signal. They are low in efficiency but give accurate and very stable control of the welding current and provide pulsing with varying waveforms and frequency.

Type 2 Switched transistorised power sources - usually DC only

Using power transistors with HF switching of the DC supply, these power sources give similar current control characteristics to Type 1, are more electrically efficient but give a smaller range of pulsing frequency and waveforms.

Type 3 Thyristor (SCR) power sources - AC/DC

Very advanced electronics, using thyristors instead of diodes on the transformer output side. These power sources give excellent current and weld time accuracy, square AC waveforms, and can be used in the pulsed mode, albeit with limited pulsing frequency response.

Type 4 AC rectifier plus inverter power sources - AC/DC

These sets are up and coming in the welding industry, very versatile, light in weight and can have many add-on features. Very cost effective and small in size with high efficiency. The response rate is generally lower than transistorised power sources, but then you can't have everything. Watch future progress in development of these undoubtedly commercially attractive sets.

The above type descriptions are in no way comprehensive, neither are they in any order of preference. Every manufacturer issues technical explanatory literature and will advise on suitability for purpose. The author does not intend to go further into technicalities but, to assist with choice (for this book only), three basic categories have been assigned to power sources to help purchasers. These are not based on any standards either national or international but are merely given to list some of the minimum features a purchaser should expect from equipment from each of the categories in addition to the basic current controls and meters:

CATEGORY A – TYPE 2

These are precision solid state, transistorised units and almost always for special and automatic use when extremely stable and consistent arcs are required. They are comprehensively specified and can be very expensive. They range down to 0.1 A up to a maximum of 100–150 A, Fig. 4.1.

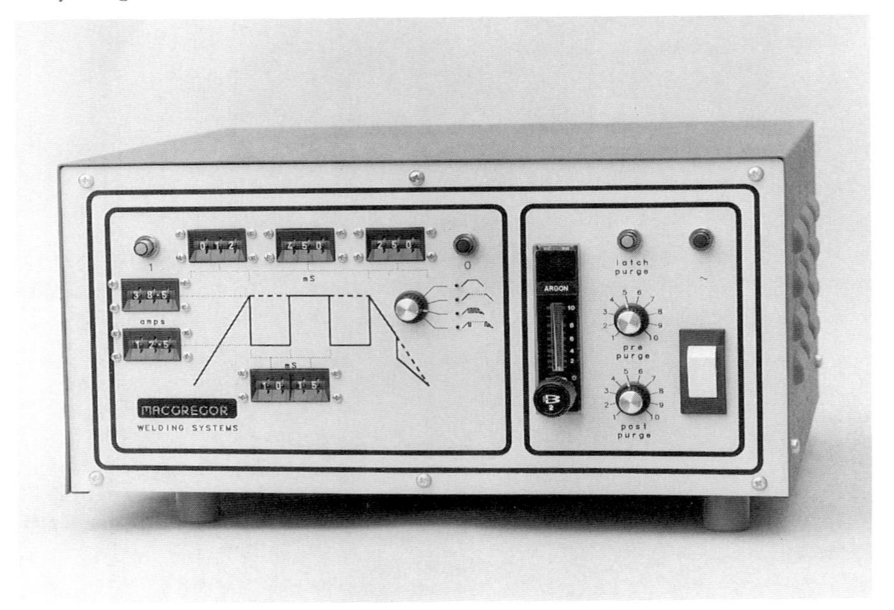

4.1 Precision solid state TIG welding power source, category A (type 2).

Features

Remote arc sequence switching;
Mains and output current stabilisation;
Arc strike by HF or similar system, *i.e.* computer friendly;
Pre-and post-purge gas flow and time controls;
Additional gas flow circuits for back purge, *etc*;
Current upslope and downslope time controls;
Pulse controls both for current levels and times;
Weld timer accurate to 0.1 s minimum;
Hand welding facility and remote current control;
Compatibility with arc length control systems, robots, computers, *etc.*

Figure 4.2 shows a welding sequence which should be available from these power sources.

Applications

Stainless and alloy edge welded bellows and diaphragms;
Transducer and load cell body welding;
Battery can top and end cap welding;

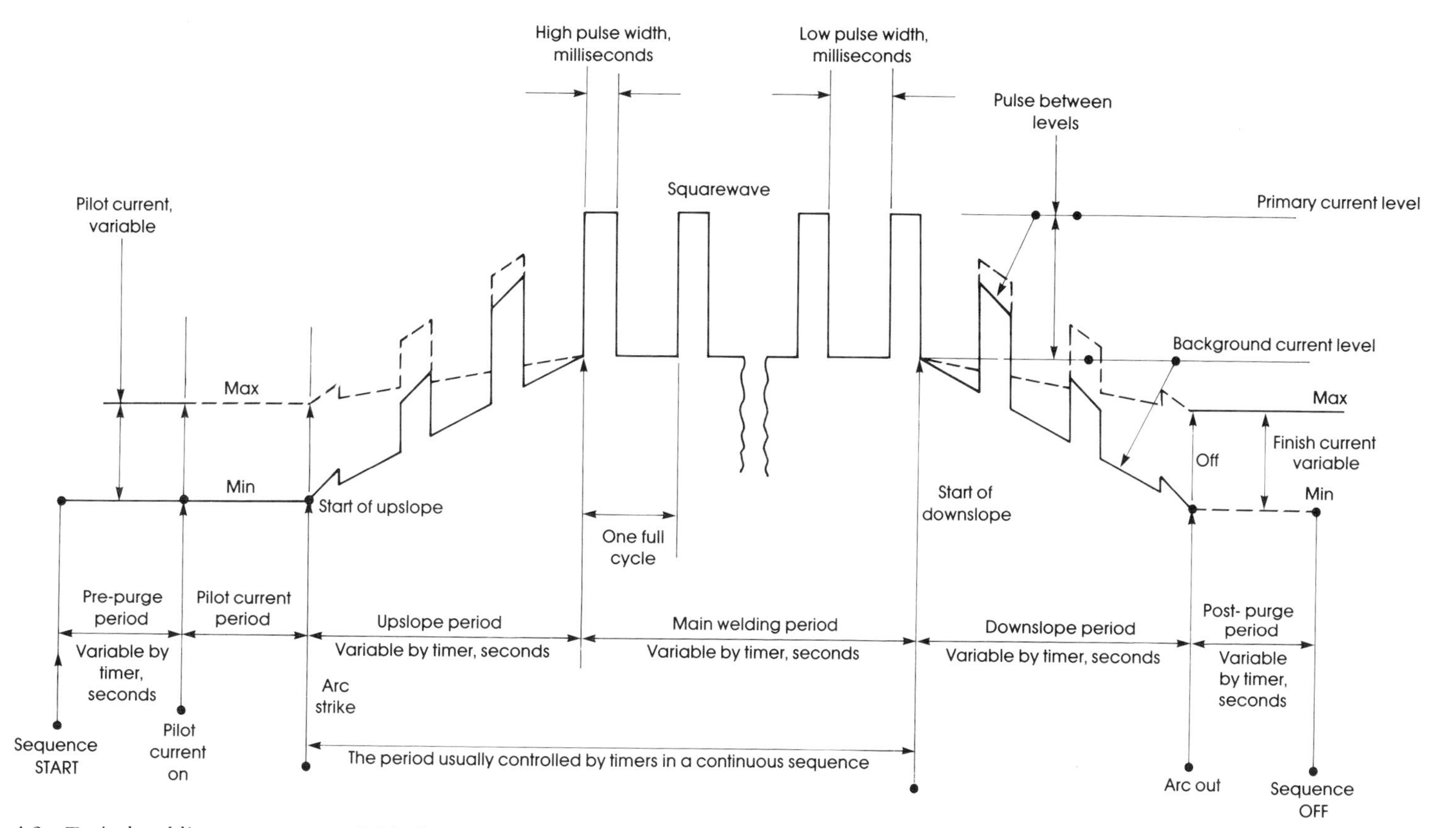

4.2 Typical welding sequence available from a category A (type 2) power source.

Mechanised and automatic welding machine use, including jewellery welding;
Operation of rotary and orbital welding heads;
Welded encapsulations of electronics;
Filter body and media assemblies in exotic metals;
Jet engine blade and seal fin repairs;
Thin section automatic aluminium welding.

CATEGORY B TYPES 1, 3 AND 4

High quality units mainly used for manual AC/DC welding. Also suitable for building in to mechanised, robotic and semi-automatic welding stations for welds of moderately high quality. Range up to a maximum of around 450 A, stable down to 8–10 A. Particularly suitable for welding aluminium and its alloys.

Features
Current level meters and controls;
Upslope and downslope time controls;
Gas flow controls;
Changeover switching, TIG to MIG/MAG;
Arc strike by HF or similar;
In-built or scope for add-on units such as pulse control, *etc*;
Pre- and post-purge gas weld timer, *etc*;
Remote arc switching.

Applications
High class workshop welding both AC and DC;
Heavy duty aluminium welding;
Edge and seam welding of canisters, air reservoirs, *etc*;
Mechanised welding machines;
Mild and stainless medium size pressure vessels;
Nuclear fit-up and repair welding.

CATEGORY C USUALLY TYPE 3

Many excellent value for money power sources come within this category and, as in category B, some can be retrofitted with pulsing and timing controls. Ideal for general and jobbing shop use. A very wide range is available and the equipment on the market is usually competitively priced. Beware of too many features, look for a good rugged build quality and service/sales availability.

Note
These Categories by no means represent the full range on the market. TIG power sources are available with capacities up to 1200 A or more and many

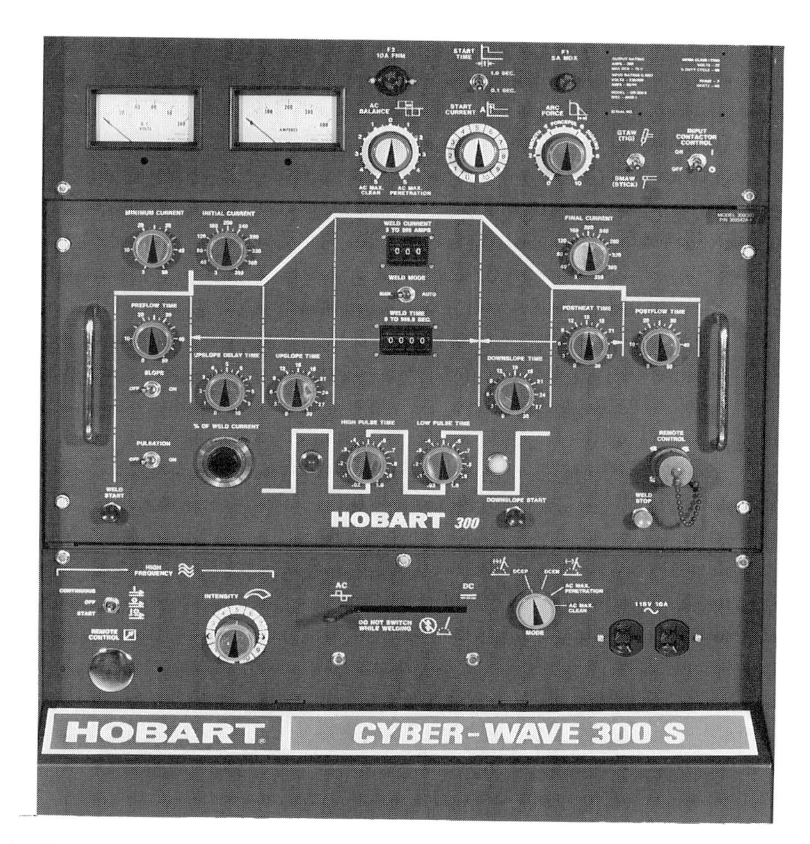

4.3 Category B TIG/MIG/MAG 300 A welding power source.

specialised machines are also on sale. A reputable manufacturer or agent should be consulted once your demands are established. If any doubt exists a professional welding society will assist you.

Installation and maintenance

Unless absolutely necessary it is best to install a power source in a convenient spot and *leave it there*, thus avoiding trailing mains supply cables, gas and water hoses, *etc*. All connections should be kept as short as possible. Make sure if you are using cylinder gas that there is adequate space near to the power source to change heavy cylinders easily and quickly (chain cylinders back to a wall if possible). Dust is prevalent in almost every welding shop and accumulates inside a power source in considerable quantities. During routine maintenance remove covers and blow out dust or remove by suction, taking care not to damage circuit boards, *etc*.

Considering the type of work it has to do a modern TIG power source can take considerable punishment and still keep working, but don't unnecessarily give it a hard life.

Calibration

Certain government, ministry, nuclear and aerospace weld procedures call for regular calibration of the welding system used, which can sometimes be a bone of contention as the meters and measuring equipment used must *themselves* be calibrated against a reference standard. Specialist companies exist to carry out checks on a contract basis as and when required and will guarantee their work.

Modern power sources are not as prone to deviate from specification as the old types so calibration and certification should only need to be carried out about once every 12 months unless the procedural contract states otherwise.

Chapter 5

Unpulsed/pulsed TIG welding

Unpulsed current AC and DC (straight)

In this mode the output current from the power source remains stable and only varies, as in *both* modes, when the arc gap is increased or decreased. Nearly all hand welding is carried out in this mode, as viewing a pulsed arc for long periods can be distressing to the operator. Figure 5.1 shows the usual welding sequence which can be expected from a good quality category B power source.

During the up and down slope periods the welding current is increased or decreased as required by the operator using a variable pedal or hand control on the torch. Operators can become extremely skilled at maintaining and repeating consistent welding parameters, keeping heat input to the required minimum. Some welds can be carried out by an expert *without*

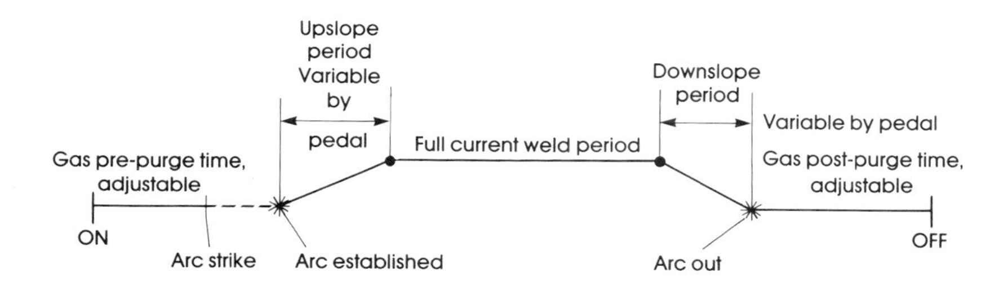

5.1 Sequence from category B power source for hand welding, with pedal control of upslope and downslope.

even using a pedal, with maximum current required set on the control panel and the operator varying the heat input by slightly increasing or decreasing the arc gap, although this cannot be done over a wide current range. The writer recommends use of a pedal current control for all manual welding as this allows full range heat variation in addition to leaving both the operator's hands free to work.

Not all power sources have controls for varying the shielding gas pre- and post-purge times, the gas being on at all times which is wasteful, so *always* try to select a set which has these functions if possible.

Metals which readily conduct, and thus lose, heat, *e.g.* copper are best welded without pulsing, as the aim is to get as much heat into the weld area as possible, particularly for thicknesses in excess of 0.5 mm.

WHY PULSE?

The advantages of pulsed DC TIG are best realised when welding metals which readily melt and flow, such as stainless steels (one possible exception being very thin, *e.g.* 0.05 mm sections and convolutes for edge welded bellows which are often better welded without pulsing). The aim of pulsing is mainly to achieve maximum penetration without excessive heat build-up, by using the high current pulse to penetrate deeply and then allowing the weld pool to dissipate *some* of the heat during a proportionately longer arc period at a lower current. Modern power sources provide a square waveform for the pulse cycle, Fig. 5.2.

In the USA the low level time B is sometimes known as the keep-alight period, an apt term. Thicker metals, say above 1.0 mm, generally require a

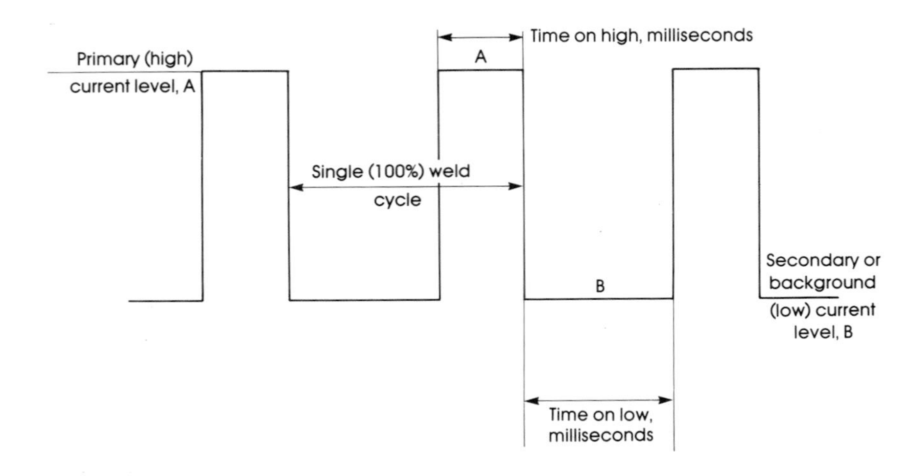

5.2 Square waveform pulse current details.

lower pulse rate than thin metals but the rate should always be what best suits a particular application. Trial and error is still the order of the day.

Pulsing can be defined as the consistent overlapping of a progressive series of spot welds. There are no rules governing pulse rate but some starting point is necessary. For stainless steel welding with a closed butt seam, a good average pulse ratio would be 1 high to 3 low: in other words A = 25% B = 75% whilst C = 66% and D = 33%, a ratio of 2 to 1. Then vary the current proportionately up or down until the required weld is achieved. It has been previously mentioned that most pulsed welding is used in automated systems as these give consistent pulse overlaps.

To achieve overlap consistency, the means of moving either the torch or the weldment along the seam must be both smooth and stepless which gives the seam a not unattractive fishscale appearance, Fig. 5.3 and 5.4.

Variations in motion cannot be tolerated particularly in precision welding as such variations increase heat input to the spots, spoiling both the strength and the cosmetic appearance of the finished weld.

One other advantage of pulsing is that the pulse action agitates/stirs the weldpool bringing impurities to the surface thus reducing inclusions and porosity.

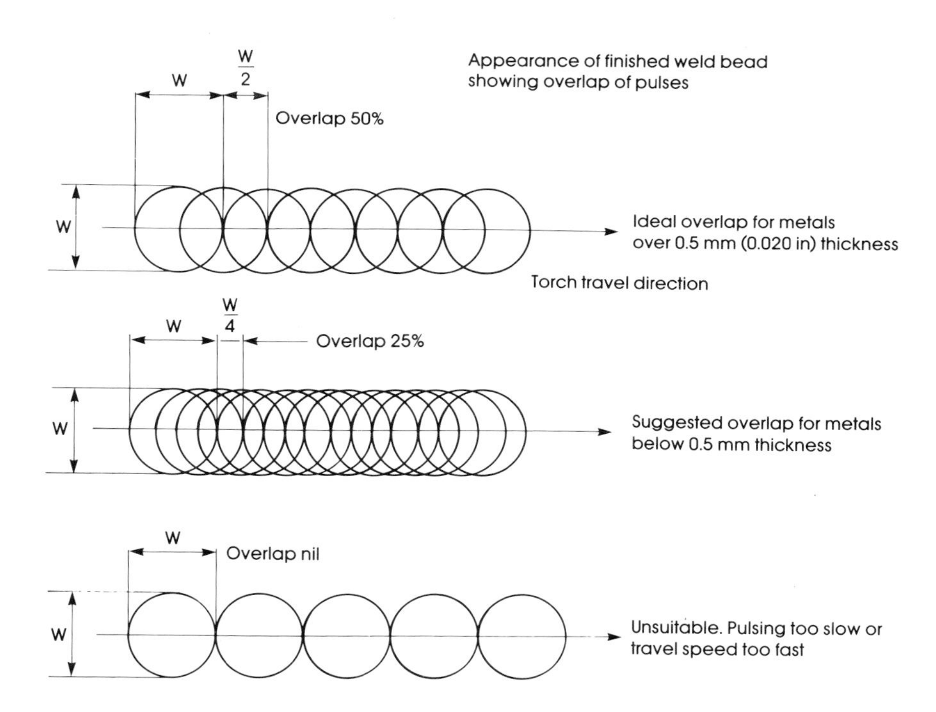

5.3 Weld spot overlap appearance.

5.4 T joint produced using pulsed TIG welding.

Average heat input to an unpulsed weld is the product of the power source dial current setting × voltage × time. To find this value for pulsed welding proceed as follows referring to Fig. 4.2.

To calculate average welding current in amps

First add high pulse time to low pulse time, both in milliseconds. This is the total weld *cycle* time (100%).

1 Divide the high pulse time by the weld cycle time which gives a high pulse figure as a decimal of 1.
2 Multiply result (**1**) by 100 = high pulse time percentage.
3 Subtract result (**2**) from 100 = low pulse time percentage.
4 Multiply primary current by (**2**)% = primary current proportion.
5 Multiply background current by (**3**)% = background current proportion.
6 Add (**4**) to (**5**) = average weld current in amps.

Then result (**6**) × arc voltage × complete weld time (sec) = heat input in joules.

EXAMPLE:

C Primary current = 16 A
D Background current = 8 A
A High pulse time = 50 millisec.
B Low pulse time = 150 millisec.

Total time taken for complete weld sequence taken as, say, 20 seconds arc time.

Arc voltage taken as 11 V

Then: $50 + 150 = 200$ *millisec* = total weld cycle time (single cycle only) or 5 complete cycles per second, i.e. 5 Hz.

Next: 1 $\frac{500}{200} = 0.25$

2 $100 \times 0.25 = 25\%$ (high pulse time, %)

3 $100 - 25 = 75\%$ (low pulse time, %)

4 $16 \times 25\% = 4.0$ A

5 $8 \times 75\% = 6.0$ A

6 $4.0 + 6.0 = 10.0$ A average current

Then $10 \text{ A} \times 11 \text{ V} \times 20 \text{ secs} = 2200$ J.

VIABILITY

Pulsed TIG welding has a drawback in that it is slower than using unpulsed current but its great advantage is that heat build-up in the component is much reduced. Indeed pulsed current is even used to weld end caps to seal the ends of small detonator cans after filling with explosive. Pulsed current also does not permit too much build-up of residual heat in circumferential or orbital welding especially where, at the end of a run, the weld seam overlaps the start.

To make the most use of arc pulsing it can be coupled to, and synchronised with, wire feed, travel speed and oscillation. This gives a more even heat spread, particularly at the seam edges of a V or J preparation and gives better fusion at these edges. Sychronisation can also be allied to arc length control and improves all the characteristics of a weld, including increased deposit rate which is most economically desirable. Many top class TIG power sources have this facility built-in to use if required.

Chapter 6

Hand welding with TIG

Manual TIG welding, like any engineering skill, needs judgement and a steady hand. The judgement is in deciding what heat input is necessary to melt the metal and achieve penetration, the steady hand guides the electrode along the seam, keeping it within the area required. Some welders attain exceptional skill and can weld very thin metals with ease, but most thin sections are best welded mechanically. Figure 6.1 shows the principles of TIG hand welding.

Preparation for welding – component

- Ensure that the correct joint preparation has been made at the weld seam;
- Remove all traces of oil or swarf from the seam edge by solvent and brushing. A brush with stainless wires or stainless wire wool must be used for mild or stainless steels. For copper, use a bronze wire brush if available;
- Wipe the seam edges with a solvent-impregnated, lint-free cloth and, important, allow to dry off thoroughly. Petroleum and other spirits are usually suitable. Check suitability of any proprietary solvents;
- Obtain the correct size and grade of filler wire or rod and degrease before using.

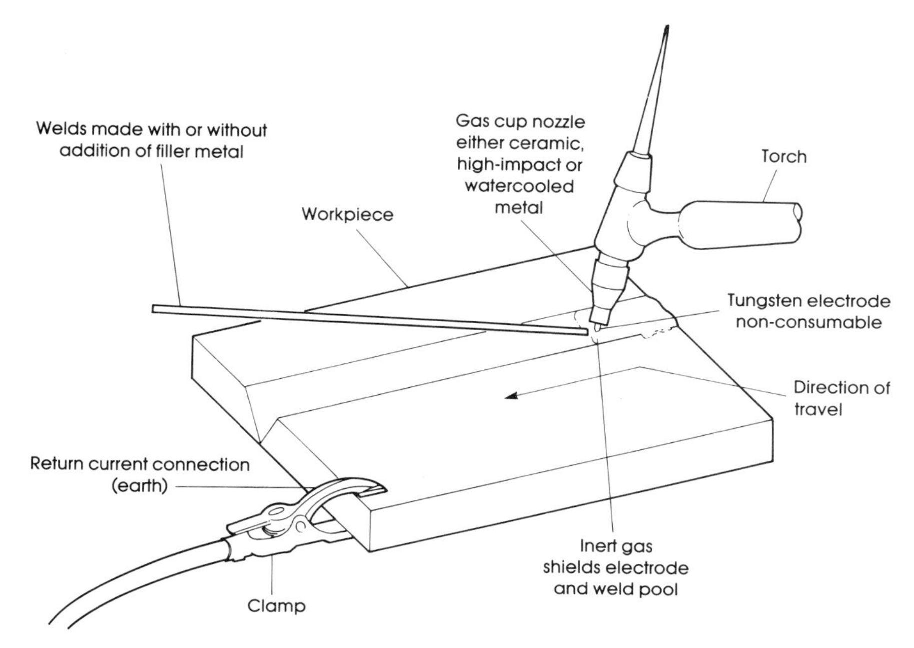

6.1 Manual TIG welding set-up.

Preparation for welding - operator

- Set the controls on the power source. It is suggested that when using a pedal current control that action be taken to set the maximum current to a safe level which cannot be exceeded, to avoid accidental burnthrough.
- Check that the correct shielding gas is available at the required flow rate. A good average amount is 7 l/min (15 ft^3/hr);
- Adjust the torch, ensuring that the correct size electrode with a suitable ground tip is fitted, electrode stickout is correct and that the type of gas cup used has the optimum orifice size;
- Clear the worktable area and/or access to the component;
- Check that the current return (earth) cable is of the correct rating and securely connected to the component being welded. This is more important than is usually considered;
- Ensure that your helmet is comfortable, fits well and that the visor screen is of sufficient density. Put on gloves and protective clothing. Check that the fume extraction equipment is working if fitted;
- Arrange for gas back purge of the seam wherever possible;
- Get in there and weld, Fig. 6.2.

The booth or welding area should be screened by curtains or similar so that the open arc cannot be seen by other personnel in the workshop.

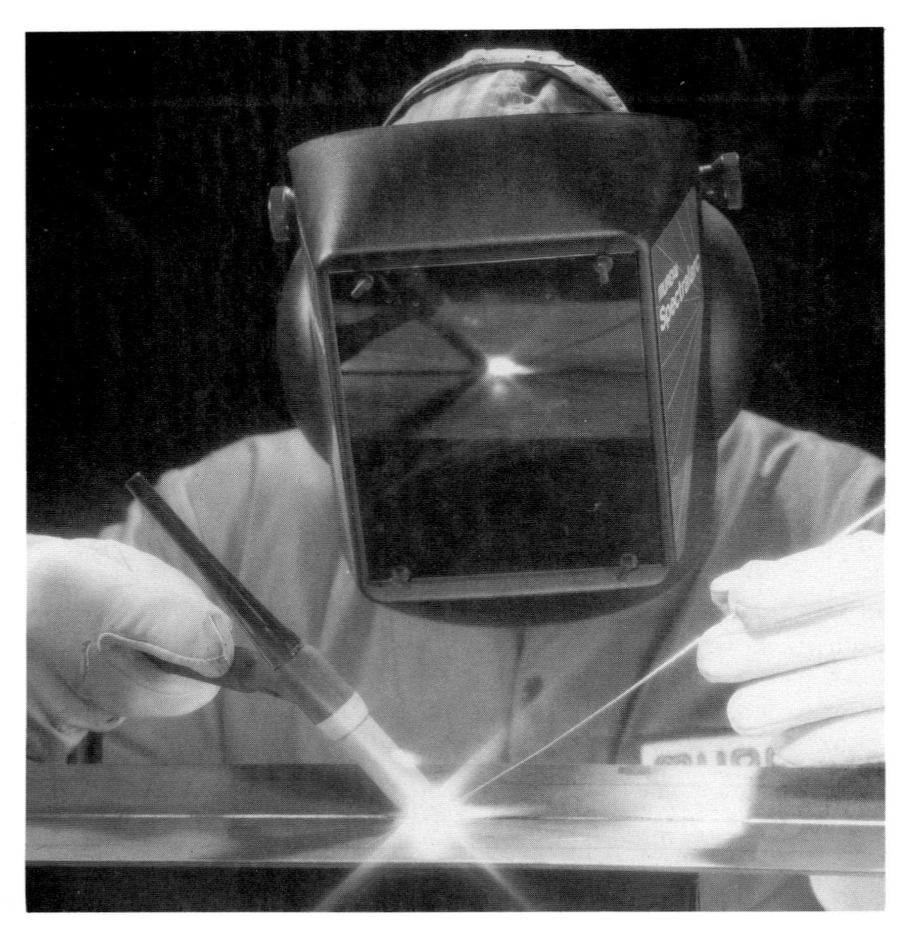

6.2 Ideal manual TIG welding conditions.

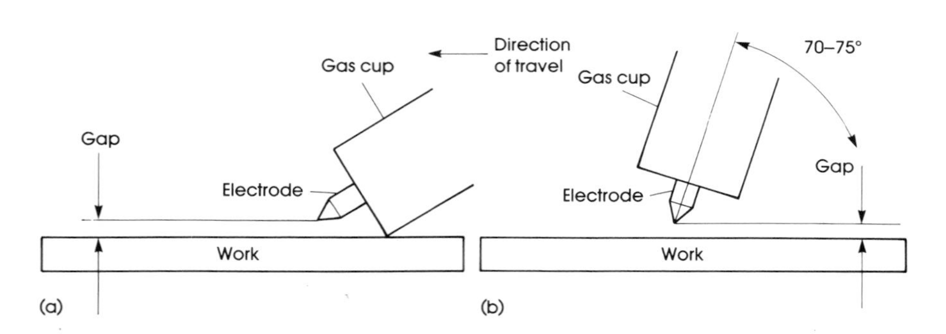

6.3 Position for: a) HF arc striking; b) Welding.

Operation

Try if possible always to weld in the flat (downhand) position and start the arc with the electrode positioned as in Fig. 6.3(a) when using HF or other open arc strike methods. A satisfactory rule of thumb when hand welding is to make the arc gap about one electrode diameter, particularly when welding aluminium. The greatest skill in TIG welding is to maintain a constant arc gap and, as far as possible, electrode-to-work angle, Fig. 6.3(b). Variations in either affect heat input and weld quality. Try never to touch-down and thus contaminate the weld pool. Always brush the weld seam after each pass to remove unwanted oxides or scale.

Weld seams

Tables 6.1 and 6.2 show a comprehensive range of butt and fillet joint preparations for stainless and alloy steels some of which can also be used for autogenous welding. It is essential that, except for some open gap root runs, the edges of the weld seam are held tightly together, perhaps by a preliminary tacking operation. Table 6.3 gives some suggested joints and procedures for spot welding. Considerable practice is needed to acquire the art of welding when using filler wire or rod. The correct position of the wire tip should be well within the shielding gas, touching the weld pool but *not* the electrode.

If it can be arranged when welding with the component on a metal topped bench, either support the gloved hand holding the torch on a smooth wooden rest or on a cool part of the work itself. Try to achieve a short weaving movement from side to side across the weld seam, closing the arc gap slightly at the extremes. This will assist by heating the metal in the proximity of the seam edges, thereby improving penetration at the seam itself. This technique is particularly useful with aluminium which requires a high heat input but can also help with any other metal.

It is advantageous and economical not to use filler wire with stainless steels of around 2.5 mm (0.100 in) thickness or less using a square butt preparation. However, the quality of most welds in carbon steels is often improved by using a compatible wire or rod as this can reduce porosity. If in doubt consult a reliable supplier who will provide wire and rod samples for test welds.

Finally, remember that practice makes perfect.

Table 6.1 Butt welding conditions for stainless and some carbon steels, DCEN, flat position

Plate and joint				Welding conditions							Consumables		
Thickness, mm	swg	Sketch	No. of passes	Nozzle bore, mm	Wire dia, mm	Electrode dia, mm	Argon flow l/min	ft^3/hr	Current, A	Speed m/hr	Wire con'n kg/m	Argon con'n l/m	Arc time min/m
0.25	33	i	1	6.4 or 9.5	—	0.8	2	4.2	8	23	—	5.2	2.6
0.35	28	i	1	6.4 or 9.5	—	0.8	2	4.2	10–12	23	—	5.2	2.6
0.56	24	i or ii	1	6.4 or 9.5	1.2	1.2	3	6.3	15–20	23–18	0.013	7.8 or 9.9	2.6 or 3.3
0.9	20	ii	1	6.4 or 9.5	1.2 or 1.6	1.2 or 1.6	3	6.3	25	15	0.015	12	4.0
1.2	18	ii	1	9.5	1.6	1.6	3	6.3	35	15	0.018	12	4.0
1.6	16	ii	1	9.5	1.6	1.6	4	8.5	50–60	12	0.022	20	5.0
2.0	14	ii	1	9.5	1.6 or 2.4	1.6	4	8.5	75	12	0.037	20	5.0
2.6	12	ii	1	9.5 or 12.7	2.4	1.6	4	8.5	85–90	9	0.045	27	6.7
3.3	10	ii or iii	1 2	9.5 or 12.7	2.4 or 3.2	1.6 or 2.4	5	10	125 90	9	0.074	67	13.4
4.8	3/16	iii	2	12.7	3.2	2.4	5	10	1st 100 2nd 125	9	0.30	67	13.4
6.4	1/4	iii	*3	12.7	3.2	2.4	5	10	1st 100 2nd 150 3rd	9	0.45	100	20.1
6.4	1/4	iv	**3	12.7	3.2	2.4	5	10	1st 125 2nd 150 3rd	9	0.30	100	20.1

* Root run – no filler

Table 6.2 Fillet welding conditions for stainless and some carbon steels, DCEN, horizontal-vertical and flat position

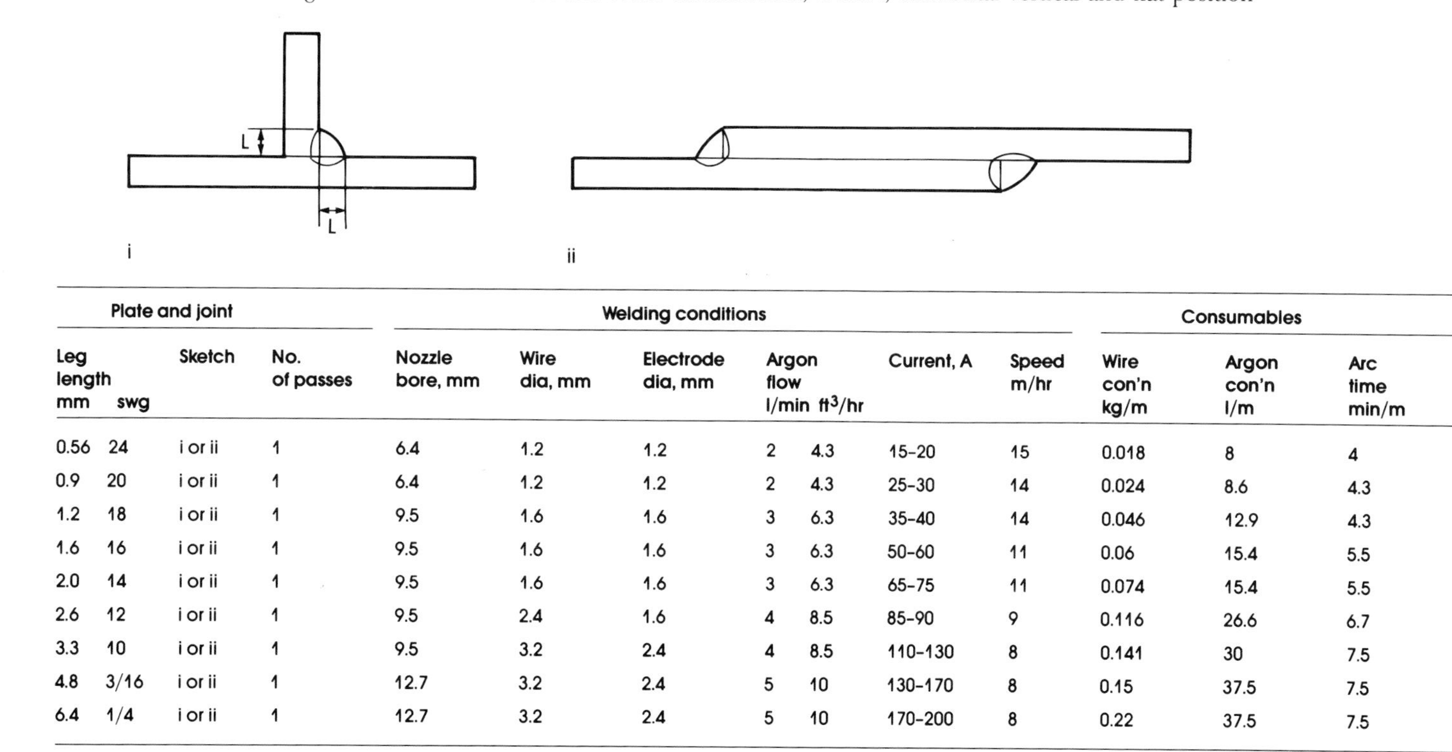

Plate and joint				Welding conditions							Consumables		
Leg length mm	swg	Sketch	No. of passes	Nozzle bore, mm	Wire dia, mm	Electrode dia, mm	Argon flow l/min	ft^3/hr	Current, A	Speed m/hr	Wire con'n kg/m	Argon con'n l/m	Arc time min/m
0.56	24	i or ii	1	6.4	1.2	1.2	2	4.3	15–20	15	0.018	8	4
0.9	20	i or ii	1	6.4	1.2	1.2	2	4.3	25–30	14	0.024	8.6	4.3
1.2	18	i or ii	1	9.5	1.6	1.6	3	6.3	35–40	14	0.046	12.9	4.3
1.6	16	i or ii	1	9.5	1.6	1.6	3	6.3	50–60	11	0.06	15.4	5.5
2.0	14	i or ii	1	9.5	1.6	1.6	3	6.3	65–75	11	0.074	15.4	5.5
2.6	12	i or ii	1	9.5	2.4	1.6	4	8.5	85–90	9	0.116	26.6	6.7
3.3	10	i or ii	1	9.5	3.2	2.4	4	8.5	110–130	8	0.141	30	7.5
4.8	3/16	i or ii	1	12.7	3.2	2.4	5	10	130–170	8	0.15	37.5	7.5
6.4	1/4	i or ii	1	12.7	3.2	2.4	5	10	170–200	8	0.22	37.5	7.5

Table 6.3 Spot welding conditions for various metals, DCEN

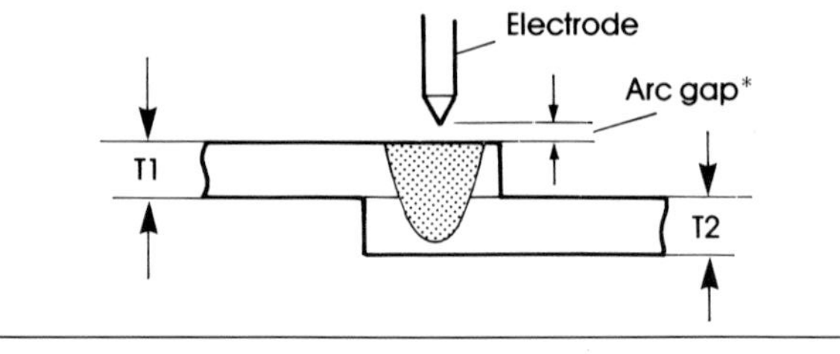

T1		Nozzle bore,	Electrode	Thickness	Welding conditions						
					swg 24	22	20	18	16	⅛	⅛
swg	mm	mm	dia, mm	T2	mm 0.56	0.71	0.91	1.21	1.62	3.2	6.4
Mild and stainless steel											
24	0.56	9.5	1.6	A (sec)†	50–55 (0.8)	55 (0.8)	55 (0.8)	55–60 (0.8)	60 (0.8)	60–65 (0.8)	65 (1.0)
22	0.71	9.5	2.4	A (sec)		75 (0.8)	75–80 (0.8)	80–85 (0.8)	85–90 (0.8)	90 (1.0)	95 (1.0)
20	0.91	12.7	2.4	A (sec)			85–90 (1.0)	85–90 (1.0)	85–90 (1.0)	95–100 (1.0)	100–105 (1.0)
18	1.21	12.7	2.4	A (sec)				140–150 (1.0)	160 (1.0)	160 (1.5)	160 (2)
16	1.62	12.7	2.4	A (sec)					175 (1.5)	180–190 (2)	200–220 (2)
Titanium and its alloys											
20	0.91	9.5	2.4	A (sec)			90 (0.5)				
18	1.21	9.5	2.4	A (sec)				140 (1.5)			
16	1.62	12.7	2.4	A (sec)					180		

Aluminium and its alloys

20	0.91	9.5	3.2	A (sec)	135 (.7)		170 (.75)	300 (2.0)
18	1.21	9.5	3.2	A (sec)		160 (1.0)		
16	1.62	12.7	3.2	A (sec)			225 (1.3)	350 (2.5)

When using downslope

T1	swg / mm	24 / 0.56	22 / 0.71	20 / 0.91	18 / 1.21	16 / 1.62
Crater time, sec		0.7	1	1.3	1.3	1.3

† These should be added to full 'arc on' times.
** Arc gap should be 0.6 mm (0.024 in) for 0.9 mm (20 swg) increasing to 0.9 mm (0.036 in) for 1.6 mm (16 swg).*
Argon flow rates should vary between 2 l/min and 6 l/min depending upon the bore of the nozzle.
The electrode should be ground to a taper of approx 1D

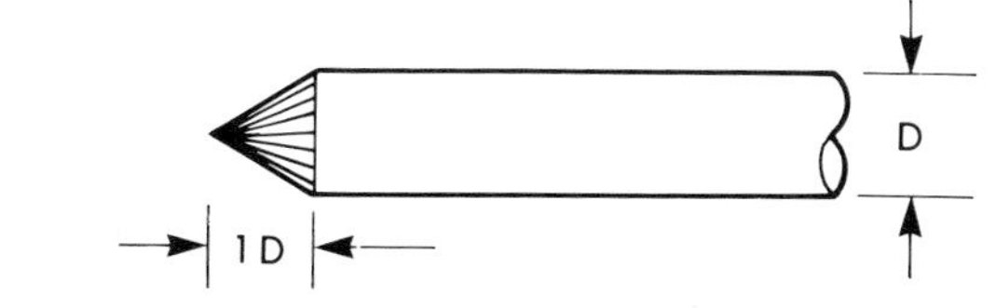

Filler wire and rods

Consumable wires or rods for TIG welding are obtainable compatible with most metals, and manufacturers will, on request, produce small batches of special materials. Table 6.4 lists a few popular proprietary filler rods sold by the main suppliers to the trade. The standards referred to are British and American, other countries have equivalents.

A list of all filler materials in wire form would more than fill this book. Specialist suppliers will advise the correct wire for any weld.

The last two joints in Table 6.5 show what to look for in a strong viable weld, the others are examples of poor welds and their causes.

Suggestions are made regarding torch and filler rod positions for four different types of joint.

BUTT JOINT

After the arc is established the torch can be raised to 70° from the work, Fig. 6.4. Establish a pool of the desired size and begin feeding filler rod into the leading edge of the pool. Hold the filler rod at about 20° as shown. Holding the arc length at about one electrode diameter, travel at a speed to produce a bead about 2 – 3 electrode diameters wide. When making a butt joint be sure to centre the weld pool on the adjoining edges. When finishing a butt weld the torch angle may be decreased to aid in filling the crater. Add enough filler metal to avoid an unfilled crater.

Cracks often begin in a crater and continue through the bead. A foot operated current control aids the finishing of a bead as current can be lowered to decrease pool size as filler metal is added.

LAP JOINT

Having established an arc, the pool is formed so that the edge of the overlapping piece and the flat surface of the second piece flow together, Fig. 6.5. Since the edge becomes molten before the flat surface, torch angle is important. The edge will also tend to burn back or undercut. This can be controlled by dipping the filler rod next to the edge as it melts away. Enough filler metal must be added to fill the joint. Finish the end of the weld the same as before. Fill the crater.

T JOINT

A similar situation exists with the T joint, Fig. 6.6, as with the lap joint. An edge and a flat surface are to be joined together. The edge again will heat up and melt sooner. The torch angle shown directs more heat on to the flat surface. The electrode may need to be extended further beyond the cup than in the previous butt and lap welds to hold a short arc. The filler rod should be dipped so it is deposited where the edge is melting away. Correct torch angle and placement of filler should avoid undercutting. Again the crater should be filled.

Table 6.4 Filler rods for manual TIG welding

Material welded	Rod composition	British standard	Nearest AWS	Comment or application
Low carbon steel	Deoxidised steel rod	BS 2901 A17 Part 1 1983	A5.288 R.60	Mild and low alloy steels
Medium carbon steel	High strength rod	BS 2901 A16	A5.2.88 R65	Also low alloy steels
Super steels	Mild steel plus some aluminium, zirconium and titanium, also silicon and manganese	BS 2901 A15	A5.18.79 ER 70S.2	Excellent for high quality root runs
Creep prone steels	Steel alloy plus 1% chromium and 0.5% molybdenum	BS 2901 A32	A5.28.79 ER 80S B.2	Boiler and superheater tubes
Creep prone steels	2% chromium, 1% molybdenum	BS 2901 A33	A5.28.79 ER 90S B.3	Consult supplier Automotive and aircraft frame tubes
Cast iron	Plus silicon	BS 1453B2		Cast iron cladding and resurfacing
Stainless steels	Rods and wires containing nickel, chromium, molybdenum and other metals in various amounts as suitable	BS 2901 part 2 347 S96 316 S92 308 S92 308 S94	A5.9.81 ER 347 ER 316L ER 308L ER 309	Consult supplier Many variations, particularly for austenitic steels
Aluminium bronze	Zinc free aluminium bronze	BS 2901 Part 3 C13	A5.777 ER-Cu A1-A2	Marine fittings
Aluminium	(a) Pure aluminium (b) + 5% silicon (c) + 10% silicon (d) + 5% magnesium (e) also with larger proportions of silicon and magnesium	BS 1453 BS 2901 Part 4 1050A 4043A 4047A 5356 5556A	A5.10.88 ER 1100 ER 4043 ER 4047 ER 5356	Use Food and aerospace Repairs General purpose Domestic and automotive AlZnMg alloys Other general purpose rods are available

Table 6.5 Weld quality inspection

	Problem	Cause
	• Excessive build-up • Poor penetration • Poor fusion at edges	Welding current too low
	• Bead too wide and flat • Undercut at edges • Excessive burn through	Welding current too high
	• Bead too small • Insufficient penetration • Ripples widely spaced	Travel speed too high
	• Bead too wide • Excessive build-up • Excessive penetration	Travel speed too low
	• Undercut • Insufficient weld deposit • Uneven penetration	Welding current too high and/or wrong placement of filler rod
	• Poor penetration • Poor fusion	Faulty joint preparation and too low welding current

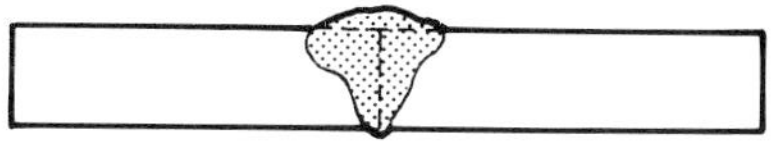

Result

- Proper build-up
- Good appearance
- Good penetration
- Bead edges fused in

Cause

Correct technique and current setting

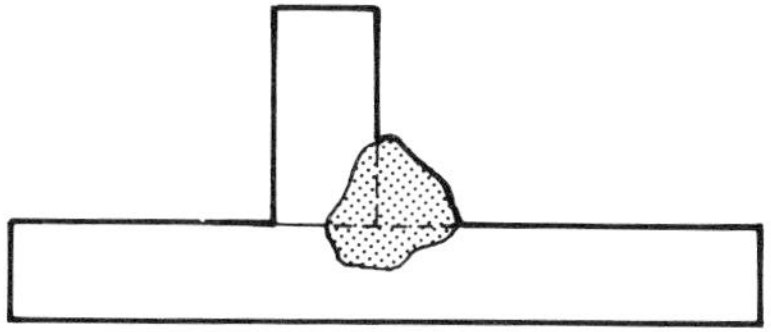

Result

- No undercut
- Legs of fillet weld equal to metal thickness
- Slightly convex bead face

Cause

Correct technique and current setting

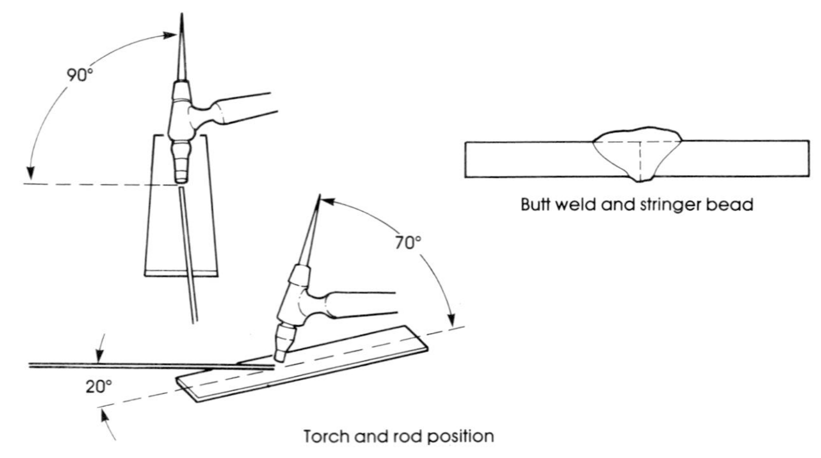

6.4 Recommended torch and filler rod angles for a butt weld and stringer bead.

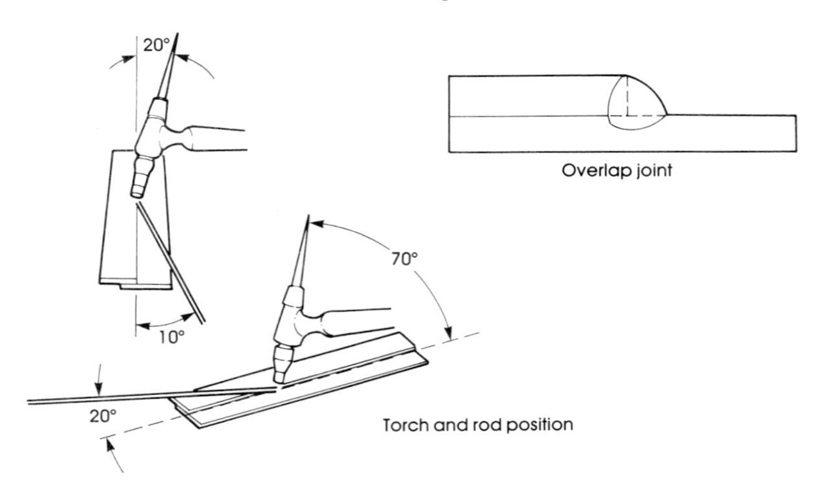

6.5 Operating angles for overlap joint.

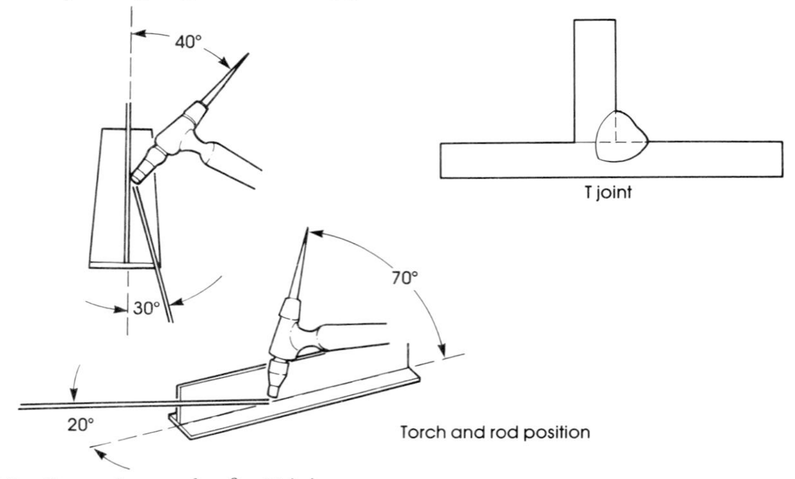

6.6 Operating angles for T joint.

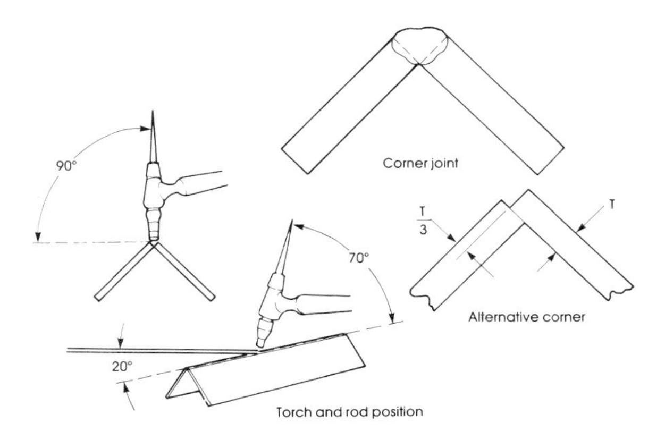

6.7 Operating angles for corner joints.

CORNER JOINTS

Correct torch and filler rod positions are shown for a corner joint, Fig. 6.7. Both edges of the adjoining pieces should be melted and the pool kept on the joint centreline. When adding filler metal, deposit sufficient to create a convex bead. A flat bead or concave deposit results in a throat thickness less than the metal thickness. On thin materials this joint design lends itself to fusion welding without filler if the fit-up is very good. An alternative corner configuration is shown for autogenous welding.

In Tables 6.6 and 6.7 suggested welding parameters, electrode sizes, filler rod sizes, gas flow rates and welding currents are given for these four types of joint in mild and stainless steels.

Table 6.6 TIG welding conditions for various joints in mild steel, DCEN

Nominal metal thickness, mm (in)	Joint type	Nominal tungsten electrode diameter, mm (in)	Filler rod diameter if required, mm (in)	Welding current, A	Argon flow, l/min (ft^3/hr)
1.5 (1/16)	Butt Lap Corner Fillet	1.5 (1/16)	1.5 (1/16)	60–70 70–90 60–70 70–90	7 (15)
3.0 (1/8)	Butt Lap Corner Fillet	1.5–2.3 (1/16–3/32)	2.3 (3/32)	80/100 90–115 80–100 90–115	7 (15)
4.5 (3/16)	Butt Lap Corner Fillet	2.3 (3/32)	3.0 (1/8)	115–135 140–165 115–135 140–170	10 (20)
6.0 (¼)	Butt Lap Corner Fillet	3.0 (⅛)	3.8 (5/32)	160–175 170–200 160–175 175–210	10 (20)

Table 6.7 TIG welding conditions for various joints in stainless steel, DCEN

Nominal metal thickness, mm (in)	Joint type	Nominal tungsten electrode diameter, mm (in)	Filler rod diameter if required, mm (in)	Welding current, A	Argon flow, l/min (ft^3/hr)
1.5 (1/16)	Butt	1.5 (1/16)	1.5 (1/16)	40–60	7 (15)
	Lap			50–70	
	Corner			40–60	
	Fillet			50–70	
3.0 (1/8)	Butt	2.3 (3/32)	2.3 (3/32)	65/85	7 (15)
	Lap			90–110	
	Corner			65–85	
	Fillet			90–110	
4.5 (3/16)	Butt	2.3 (3/32)	3.0 (1/8)	100–125	10 (20)
	Lap			125–150	
	Corner			100–125	
	Fillet			125–150	
6.0 (¼)	Butt	3.0 (⅛)	3.8 (5/32)	135–160	10 (20)
	Lap			160–180	
	Corner			135–160	
	Fillet			160–180	

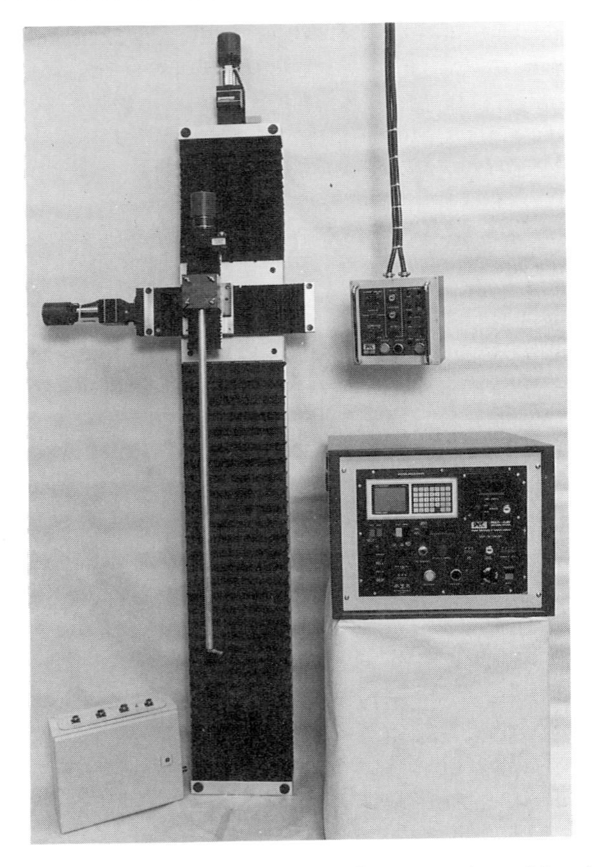

6.8 Mechanised internal TIG cladding torch and X, Y, Z axis positional controller.

TIG cladding and hardsurfacing

Coating the surface of one metal with a thin layer of another to provide wear or corrosion resistance has been carried out by TIG welding for many years. Expensive parts can often be rebuilt as required and then machined back to size. Tanks for corrosive liquids can be surfaced internally and valves in oil and chemical pipelines rebuilt. The reclamation of worn components is very cost effective.

Coating processes are mostly carried out by hand but much development is taking place in mechanical and automatic methods, Fig. 6.8.

Four types of internal cladding that can be carried out by programmed TIG are shown in Fig. 6.9 and turnkey equipment can be purchased for this purpose.

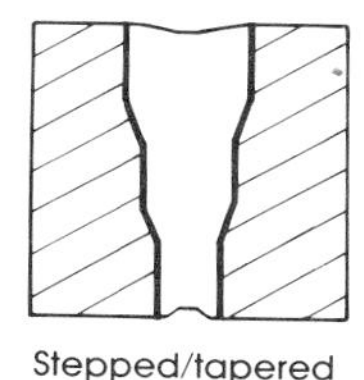

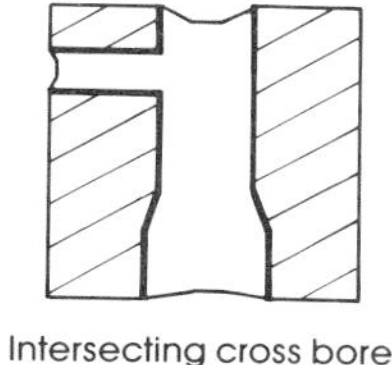

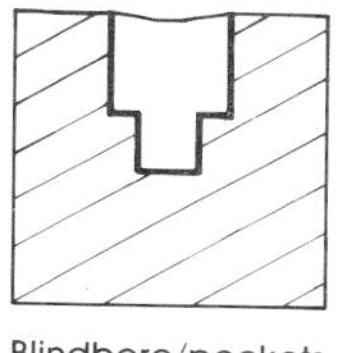

6.9 Various types of internally clad surfaces.

COMPONENT GEOMETRIES FOR INTERNAL CLADDING

- Plain, stepped and tapered bores;
- Bores with intersecting cross bores;
- Conical and spherical surfaces, internal and external, *e.g.* pressure vessel nozzles, control valve bores, seal faces and ball valves;
- Well head bonnets, gate valves;
- Components with pockets or blind bores;
- Valve ring grooves;
- Flat plates (gates) with/without intersecting holes.

The size of bores that can be clad varies from 25 mm upwards in diameter and up to 2000 mm length. The size and geometry of the bore dictate the welding process used. For plain, parallel bores up to 80 mm diameter the cold wire TIG process is used to weld the first pass, with hot wire TIG for subsequent passes. This combination provides good fusion/dilution properties and high deposition rates. Larger diameters are welded using hot wire TIG (or synergic MIG). The ultimate choice of process is based on economics and the weld properties required.

POINTS TO REMEMBER

- Ensure that metal surfaces are clean and correctly prepared;
- Use the correct wire or rod, many types are available;
- Keep the layer as consistent and as thin as practicable. Too thick a layer can be worse than one that is too thin;
- Preheat the component before commencing work if possible;
- Keep heat input as consistent as possible for a homogeneous layer.

For some stainless base materials the cladding process can give rise to severe hydrogen cracking problems, particularly in the bead overlap areas. Alloys with more than 0.3% carbon are troublesome but use of Inconel alloys can assist. Consult a specialist supplier dealing in surfacing wire and rod. TIG is a rather slow process for cladding, but has advantages such as arc concentration and heat control. Pulsed TIG helps in reducing porosity but slows the process even further although improving quality.

Chapter 7

Machine TIG welding

Autogenous mode

Of all the welding processes TIG, by reason of its cleanness and controllability, is probably the most suitable for machine and automatic use. Some most ingenious and economical solutions have been found for producing consistent automatic welds of full strength and good cosmetic appearance. For reasons of clarity we now subdivide the various mechanical categories A, B and C as follows:

A Semi-automatic;
B Fully automatic, sometimes with computer programming;
C Automatic loading and unloading of the component added to A.

CATEGORY A

These are machines where loading is by hand and the linear or orbital motion and the welding process are brought on stream by the operator in sequence as required. A typical sequence would be load, clamp, start motion, weld on, weld off, stop motion, unclamp, unload, where each step in the sequence is manually initiated. This allows the operator to control each item individually in the sequence and is ideal for a budget type low production system.

CATEGORY B

A logical first step from category A. In these systems loading and unloading are still carried out by hand, after which one button can be pressed and the clamp, start motion, weld on, weld off, stop motion, unclamp items are automatically switched in as a continuous sequence. Many modern TIG power sources (category A, chapter 4) have remote sockets fitted to facilitate sequenced weld switching. A typical one button sequence would then be possible, *e.g.* clamp, start motion, pre-purge on, arc strike, current upslope, full weld on, full weld off, current downslope, arc out, stop motion, post-purge, gas off, unclamp, with loading and unloading still by hand. This is adequate for medium production and is often preferred for high precision welding as each finished weldment can be inspected by the operator before repeating the process.

Clamping is usually pneumatic and other items such as wire feeders, arc length control systems, safety screens and additional gas purge systems can be added to the system as and when required. Figure 7.1 shows a typical system with many of the functions of this category controlled by a single button.

7.1 Manually loaded automatic TIG welding lathe with single button sequence control.

CATEGORY C

As a further and equally logical step, these machines can be fitted with bowl or linear feeders and can often attain quite high production rates. It is usual on such machines to incorporate fail safe warning devices such as no gas, or component not in position, to reduce rejects. At the time of writing such machines are not common but are becoming increasingly so as techniques progress.

Automatic and semi-automatic welding machines are often built to machine tool accuracy. For welding thin sections the arc tip must be very accurately positioned and a constant arc gap maintained. Rotational and linear motion must be variable and, once started, capable of being maintained to an accuracy of ± 1% of set speed with instant stop and start functions. It is nearly always best to move the component rather than the arc and piece parts should be made to a high standard of finish and accuracy. For circumferential welds in particular, ensure good concentricity. Cleanness of the components is, of course, mandatory. Some advice on heat sinks for automatic welding will be given later along with suitable joint designs.

Welding parameters for automatic machines

Category A (chapter 4) power sources usually have all the necessary meters and dials to achieve a staggering number of different and very accurate combinations of arc time, pulse rate and current levels, none of which is any use if the business end, *i.e.* the torch and electrode, is not up to the job. For example, a greater accuracy of electrode tip is needed if fine welds are to be achieved and this is where a tip grinder really becomes an asset. There now follows some general advice on several very important points to consider when designing or using a machine welding system:

- Use the correct size and material for electrode. Keep point clean and sharp and keep stickout to a minimum, say 1.5 mm (1/16 in) beyond the tip of the ceramic shroud if accessibility to the weld allows. Otherwise, as short as possible;
- Ensure a warm, dry supply of the correct inert gas at a suitable flow rate. Do not confuse velocity with volume particularly when using small diameter gas cups. What *is* important is the spread of gas, and gas lenses fitted inside the shroud help with this by diffusing the gas flow. Too much gas can be as bad as too little; always back-purge the underside of the weld seam whenever possible.
- Try first a 2:1 primary to background current ratio and a 1:4 high to low pulse ratio;

- By experimentation select correct weld speed in automatic applications and adjust these ratios if necessary;
- Set the smallest practicable arc gap ± 0.05 mm (0.002 in) and maintain;
- Ensure a good fit-up and heat balance at the weld point (advice on this is given in a following chapter);
- Keep the weld area clean and uncontaminated. Aluminium for example oxidises very quickly: weld within ten minutes of cleaning;
- Provide current return (or earthing) sufficient to carry the *full* welding current. This is most important and often overlooked. Never use plain carbon brushes with automatic fixtures. Copper and brass are more suitable and copper-carbon is suitable below 50 A average current. A spring-loaded shoe should be in contact with a moving part of the jig as near as possible to the weldment. As a guide, using copper brushes, there should be *at least* 600 mm² (1 in²) of brush surface in contact per 50 A of welding current (or pro rata). It is also useful to coat the rubbing surfaces lightly with fine graphite powder when running in a shoe.
- Try for a 5 - 10 per cent overrun of a circumferential or linear weld seam allowing upslope and downslope to occur whilst the component is still being moved under the arc.

WELDING CURRENT SELECTION AND TRAVEL SPEED

DCEN welding – for thin section autogenous welds only

The correct setting of welding current values can only be determined by experiment and production of samples, but the following is a guide for a safe starting point. Note that everything depends on smooth consistent travel either of the arc along the seam or vice versa. The values are given per 0.025 mm (0.001 in) of metal thickness and for a square butt joint.

- Carbon, mild and stainless steels – 0.50 A
- Aluminium and alloys – 1.5 A
- Copper and alloys – 2–3 A

These are *average* welding currents and are purposely given low to avoid ruining too many expensive samples.

Regarding travel speeds these will usually range from 2 - 6 mm/sec dependent on material and thickness, although in production much higher speeds have been attained.

Circumferential welds

A welder's lathe is a great asset for coping with a wide range of cylindrical welded components. Heat sinks, sometimes watercooled, can be fixed to the headstock and tailstock with pneumatic positioning and clamping units fixed to and adjustable along the lathe baseplate, Fig. 7.2.

Much can also be accomplished using a fairly simple welder's turntable in the horizontal position with a rigid adjustable device for positioning the torch. It is essential that any rotating equipment is mounted on a firm heavy table or base, sometimes placing the power source within the frame to provide instant access to the controls plus additional stability.

7.2 Precision mini-lathe used in aerospace component welding. Manual loading, automatic weld sequencing.

Orbital welding

This is defined as the condition in which the arc travels around the external periphery of a fixed pipe or tube weld seam. Much ingenuity has been applied, particularly in the USA, to designing and producing orbital welding heads and several American manufacturers offer extensive standard ranges of such equipment. The USA seems to be the home of automated orbital tube welding. In addition this is one area to which computer programmed power sources have been extensively applied where a large number of programmes can be stored and called up as required. It is possible to purchase a compact and portable package system, Fig. 7.3, which operates over a large range of pipe and tube outer diameters, with many varying welding parameters to give choice of welding position.

For tubes up to 200 mm (8 in) outer diameter most precision orbital heads are the totally enclosed type where autogenous welding takes place within the body of the head in a fully gas purged atmosphere. These heads can be operated in conditions of limited access as they are comparatively small in size.

For large pipes a full function in position (FFIP) mechanised welding head can be a solution. These units, whilst very expensive, can work on a large range of pipe sizes often from 50 mm (2 in) up to 2000 mm (80 in). They also allow for several passes and filler runs for heavy wall sections. These units usually have all or some of the following facilities: arc length control, wire feed, cross seam arc oscillation (weaving) and various pulsing functions. They normally run on a split geared ring clamped around the pipe being welded. The fact that one ring - one pipe size is the norm accounts for much of the high cost but in nuclear construction one weld can be critical, in which case the advantages far outweigh the high cost of the ring, Fig. 7.4.

FFIP weld heads can be programmed to work over a wide range of weld parameters and, very importantly, with the electrode in a wide range of positions, from 12 o'clock to 6 o'clock and all positions in between. This facility is essential for fixed pipes whereas rotating pipes are usually welded in the 12 o'clock position.

Table 7.1 lists the available schedule pipe sizes in their nominal diameters and is included here to show what sizes can be autogenously welded using enclosed orbital welding heads. However, such information is often difficult to find when required unless you are working in the pipe fitting and oil industries so its inclusion here is justified if only for the purpose of convenience.

Internal bore welding

This is a mechanised process where the electrode is rotated around the weld seam within the tube bore. This method requires very accurate arc positioning, smooth rotation, low, accurate gas flow and suitable positional weld programming. As it is not possible to guarantee the smoothness of the weld bead surface or porosity content it is not often used for food processing pipes, being mainly confined to boiler tubes and heat exchangers where it has been successfully applied to welds more than 1800 mm (6 ft) down from the tube end. Most welding systems are custom built to suit a particular weld, Fig. 7.5 and 7.6.

Internal bore welding, because of its application within a confined space, is almost always carried out using pulsed TIG in the DCEN mode and a square butt preparation.

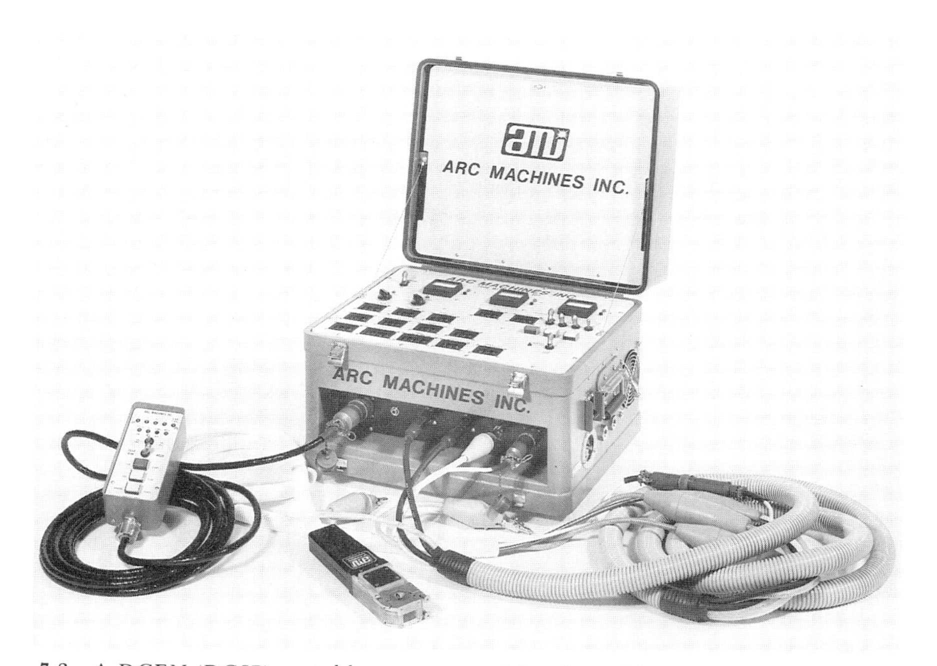

7.3 A DCEN (DCSP) portable, programmable tube welding system.

7.4 Full function orbital pipe welding head equipped with leading and trailing edge video viewing for hazardous location welding.

Table 7.1 Schedule pipe sizes to ANSI B 36.10 specification

Nom pipe size, in	Outside diameter, mm	Outside diameter, in	Schedule 5 mm	Schedule 5 in	Schedule 10 mm	Schedule 10 in	Schedule 20 mm	Schedule 20 in	Schedule 30 mm	Schedule 30 in	Standard, STD mm	Standard, STD in	Schedule 40 mm	Schedule 40 in	Schedule 60 mm	Schedule 60 in	Extra strong XS mm	Extra strong XS in	Schedule 80 mm	Schedule 80 in	Schedule 100 mm	Schedule 100 in	Schedule 120 mm	Schedule 120 in	Schedule 140 mm	Schedule 140 in	Schedule 160 mm	Schedule 160 in	Extra strong XXS mm	Extra strong XXS in
⅛	10.3	0.405	—	—	1.23	0.049	—	—	—	—	1.73	0.068	Identical		—	—	2.41	0.095	Identical		—	—	—	—	—	—	—	—	—	—
¼	13.7	0.540	—	—	1.65	0.065	—	—	—	—	2.24	0.088	to		—	—	3.02	0.119	to		—	—	—	—	—	—	—	—	—	—
⅜	17.1	0.675	—	—	1.65	0.065	—	—	—	—	2.31	0.091	standard		—	—	3.20	0.126	extra		—	—	—	—	—	—	—	—	—	—
½	21.3	0.840	1.65	0.065	2.11	0.083	—	—	—	—	2.77	0.109			—	—	3.73	0.147	strong		—	—	—	—	—	—	4.78	0.188	7.47	0.294
¾	26.7	1.050	1.65	0.065	2.11	0.083	—	—	—	—	2.87	0.113			—	—	3.91	0.154			—	—	—	—	—	—	5.56	0.219	7.82	0.308
1	33.4	1.315	1.65	0.065	2.77	0.109	—	—	—	—	3.38	0.133			—	—	4.55	0.179			—	—	—	—	—	—	6.35	0.250	9.09	0.358
1¼	42.2	1.660	1.65	0.065	2.77	0.109	—	—	—	—	3.56	0.140			—	—	4.85	0.191			—	—	—	—	—	—	6.35	0.250	9.70	0.382
1½	48.3	1.900	1.65	0.065	2.77	0.109	—	—	—	—	3.68	0.145			—	—	5.08	0.200			—	—	—	—	—	—	7.14	0.281	10.16	0.400
2	60.3	2.375	1.65	0.065	2.77	0.109	—	—	—	—	3.91	0.154			—	—	5.54	0.218			—	—	—	—	—	—	8.74	0.344	11.07	0.436
2½	73.0	2.875	2.11	0.083	3.05	0.120	—	—	—	—	5.16	0.203			—	—	7.01	0.276			—	—	—	—	—	—	9.53	0.375	14.02	0.552
3	88.9	3.500	2.11	0.083	3.05	0.120	—	—	—	—	5.49	0.216			—	—	7.62	0.300			—	—	—	—	—	—	11.13	0.438	15.24	0.600
3½	101.6	4.000	2.11	0.083	3.05	0.120	—	—	—	—	5.74	0.226			—	—	8.08	0.318			—	—	—	—	—	—	—	—	16.15	0.636
4	114.3	4.500	2.11	0.083	3.05	0.120	—	—	—	—	6.02	0.237			—	—	8.56	0.337			—	—	11.13	0.438	—	—	13.49	0.531	17.12	0.674
5	141.3	5.563	2.77	0.109	3.40	0.134	—	—	—	—	6.55	0.258			—	—	9.53	0.375			—	—	12.70	0.500	—	—	15.88	0.625	19.05	0.750
6	168.3	6.625	2.77	0.109	3.40	0.134	—	—	—	—	7.11	0.280			—	—	10.97	0.432			—	—	14.27	0.562	—	—	18.26	0.719	21.95	0.864
8	219.1	8.625	2.77	0.109	3.76	0.148	6.35	0.250	7.04	0.277	8.18	0.322			10.31	0.406	12.70	0.500			15.09	0.594	18.26	0.719	20.62	0.812	23.01	0.906	22.23	0.875
10	273.1	10.750	3.40	0.134	4.19	0.165	6.35	0.250	7.80	0.307	9.27	0.365			12.70	0.500	12.70	0.500	15.09	0.594	18.26	0.719	21.44	0.844	25.40	1.000	28.58	1.125	25.40	1.000
12	323.9	12.750	4.19	0.165	4.57	0.180	6.35	0.250	8.38	0.330	9.53	0.375	10.31	0.406	14.27	0.562	12.70	0.500	17.48	0.688	21.44	0.844	25.40	1.000	28.58	1.125	33.32	1.312	25.40	1.000
14	355.6	14.000	—	—	6.35	0.250	7.92	0.312	9.53	0.375	9.53	0.375	11.13	0.438	15.09	0.594	12.70	0.500	19.05	0.750	23.83	0.938	27.79	1.094	31.75	1.250	35.71	1.406	—	—
16	406.4	16.000	—	—	6.35	0.250	7.92	0.312	9.53	0.375	9.53	0.375	12.70	0.500	16.66	0.656	12.70	0.500	21.44	0.844	26.19	1.031	30.96	1.219	36.53	1.438	40.49	1.594	—	—
18	457.2	18.000	—	—	6.35	0.250	7.92	0.312	11.13	0.438	9.53	0.375	14.27	0.562	19.05	0.750	12.70	0.500	23.83	0.938	29.36	1.156	34.93	1.375	39.67	1.562	45.24	1.781	—	—
20	508.0	20.000	—	—	6.35	0.250	9.53	0.375	12.70	0.500	9.53	0.375	15.09	0.594	20.62	0.812	12.70	0.500	26.19	1.031	32.54	1.281	38.10	1.500	44.45	1.750	50.01	1.969	—	—
22	558.8	22.000	—	—	6.35	0.250	9.53	0.375	12.70	0.500	9.53	0.375	—	—	22.23	0.875	12.70	0.500	28.58	1.125	34.93	1.375	41.28	1.625	47.63	1.875	53.98	2.125	—	—
24	609.6	24.000	—	—	9.53	0.250	9.53	0.375	14.27	0.562	9.53	0.375	17.48	0.688	24.61	0.969	12.70	0.500	30.96	1.219	38.89	1.531	46.02	1.812	52.37	2.062	59.54	2.344	—	—
26	660.4	26.000	—	—	7.92	0.312	12.70	0.500	—	—	9.53	0.375	—	—	—	—	12.70	0.500	—	—	—	—	—	—	—	—	—	—	—	—
28	711.2	28.000	—	—	7.92	0.312	12.70	0.500	15.88	0.625	9.53	0.375	—	—	—	—	12.70	0.500	—	—	—	—	—	—	—	—	—	—	—	—
30	762.0	30.000	—	—	7.92	0.312	12.70	0.500	15.88	0.625	9.53	0.375	—	—	—	—	12.70	0.500	—	—	—	—	—	—	—	—	—	—	—	—
32	812.8	32.000	—	—	7.92	0.312	12.70	0.500	15.88	0.625	9.53	0.375	17.48	0.688	—	—	12.70	0.500	—	—	—	—	—	—	—	—	—	—	—	—
34	863.6	34.000	—	—	7.92	0.312	12.70	0.500	15.88	0.625	9.53	0.375	17.48	0.688	—	—	12.70	0.500	—	—	—	—	—	—	—	—	—	—	—	—
36	914.4	36.000	—	—	7.92	0.312	12.70	0.500	15.88	0.625	9.53	0.375	19.05	0.750	—	—	12.70	0.500	—	—	—	—	—	—	—	—	—	—	—	—

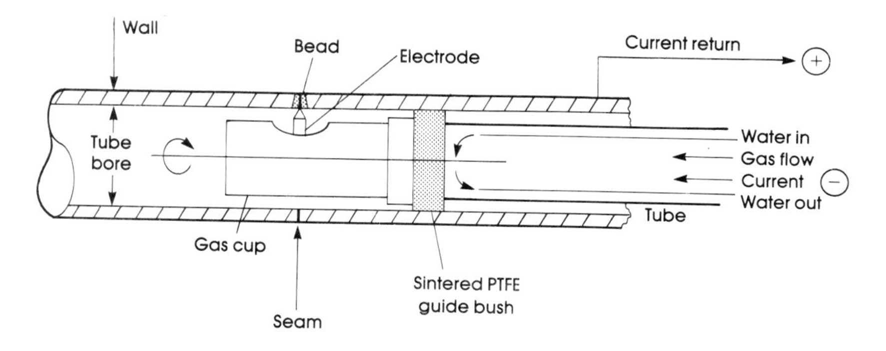

7.5 Details of custom built internal bore welding torch.

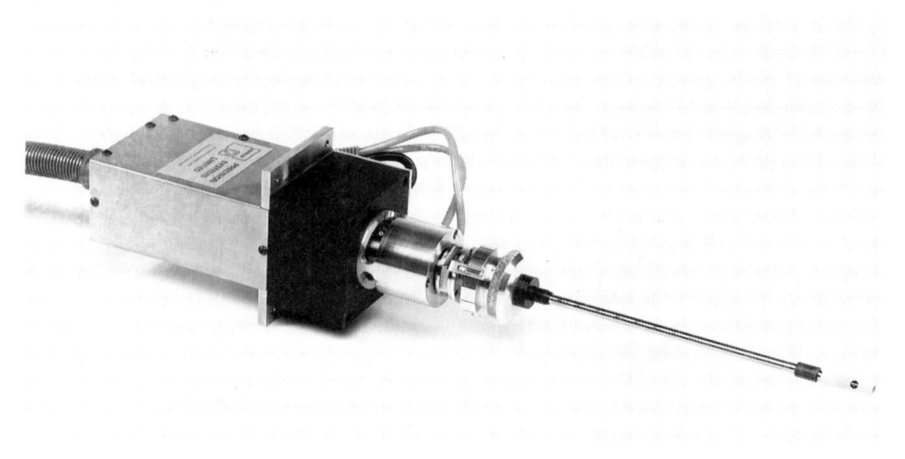

7.6 Special long reach head for internally welding 16mm bore heat exchanger tubes.

Summary

Weld mechanisation has many advantages, one of which is removal of the human element. However, in most cases cost is usually the greatest consideration when deciding to mechanise or not.

One ingenious type of orbital welding head is for production of reverse bends in boiler and heat exchanger tubing, Fig. 7.7.

Heads must be of minimum size to fit in the available gap and at the same time be heat resistant enough to withstand a large number of welds without excessive heat build-up. Welds of this type are traditionally done by hand, and as yet not all reverse bends can be mechanically welded because of lack of clearance for the welding head.

Another welding head carries out welds of tubes to tube plates (or tube sheets) in boilermaking, Fig. 7.8. These heads are often located in one tube end whilst welding another and some possible joint designs and weld configurations are shown later.

These heads are usually coupled to a sophisticated power source and are a prime choice for computer programming as a large number of similar welds are often made during a shift.

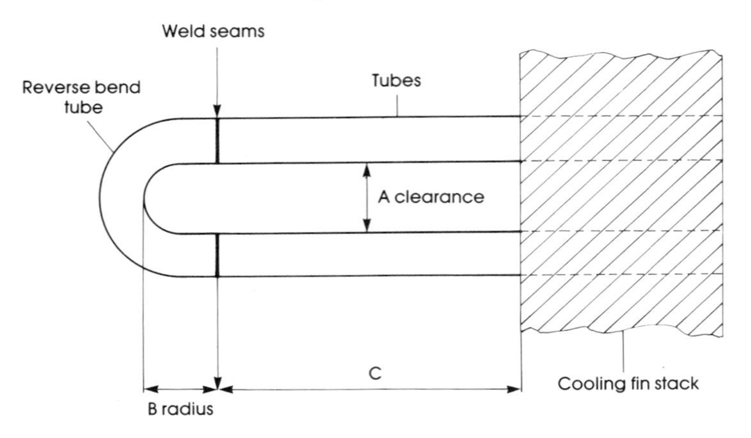

7.7 Reverse bend heat exchanger tubes with limited access at A, B and C.

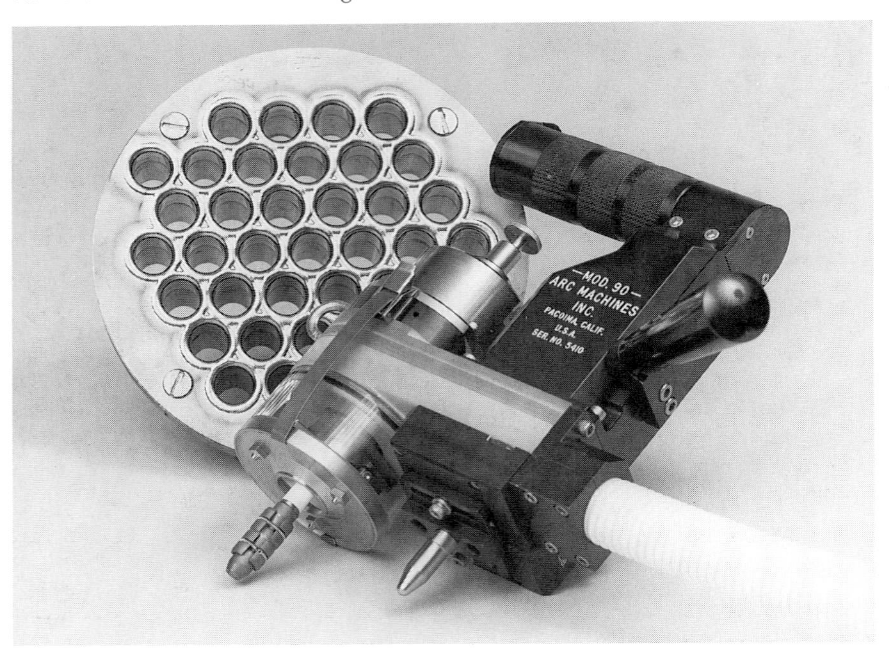

7.8 Tube to tube sheet fusion-only welding head.

Table 7.2 Welding conditions and procedure record sheet

Company:	Compiled by:	Date:	
Component:	Material details:		
Power source used:			
Arc length control used: Yes/No	Arc gap:		
If yes enter:	ALC delay period	Voltage setting	
Other ALC control settings			
ELECTRODE	Type: Size:		

Electrode dia 'D'

Point grinding

Included angle

DC

90°

Included angle

$\frac{D}{2}$

AC

WELDING CURRENT	Pulsed/unpulsed AC/DC	Arc strike current level A	
Upslope period sec		Downslope period sec	
Peak current A		Background current A	
Peak pulse time msec		Background current time msec	
Pulses per sec		% pulse time	
Time on full weld current sec	Total weld time sec	Electrode stick-out distance mm	
SHIELDING GASES	Type to electrode	Type to back purge	
Gas cup type	Flow rate l/min	Flow rate l/min	
Gas cup orifice	Gas lens used Yes/No	Pre-purge time sec	Post-purge time sec
FIXTURE	Rotation speed, rpm	Traverse speed mm/sec	Weld seam overlap mm

Other data and comments

WELD PROCEDURE RECORDS

As welding becomes more and more technical and parameters more complex, it becomes necessary to keep accurate records so that processes can be exactly repeated later. Even if the procedures have been fed on to tape or into computer memory these can sometimes become destroyed or corrupted, so that when a weld sequence is in preparation it is prudent to make out a written parameter sheet. A suggested design for such a sheet is shown in Table 7.2. The author has used this type of record sheet for many years with great success. The sheet is designed for mechanical or automatic welding; a similar but less complicated record should be kept for hand welding.

Chapter 8

Computer and microprocessor control of TIG welding

Among welding processes TIG is probably the most suitable for remote programming. The previous chapter has given some details of mechanical automation, this chapter very briefly covers the subject of use of computers and microprocessors for control of individual items in systems using an IBM PC or other similar equipment. This control involves ensuring that the power source is in the correct mode, welding current and arc voltage are displayed by instrumentation, weld finished commands given, *etc*.

Microprocessors are extensively used:

- In the power source;
- For transmitting data to external peripheral equipment;
- To interface with dedicated peripheral equipment;
- For programming constraints.

The advantage of microprocessor control is that accurate and repeatable arcing performance with minimal user involvement is made possible. The range encompassed can be from ordinary mechanisation to fully automatic operation of the welding station, control of turntables, traverses, and wire feed units.

Power source

The microprocessor controlled power source controls the following functions:

- Pre-purge gas flow volume and time;
- Initiate and establish welding current commands;
- Upslope, pulsing and downslope;
- Peak and base level welding current;
- Post-purge gas flow volume and time;
- Welding commands to external equipment.

Standard microprocessors are of the 8-bit variety, *e.g.* Z80, M6800, *etc*, but 16-bit and 4-bit types may also be used. The type reflects the complexity of the functions and resolution required in time and level.

The nomenclature of 4, 8 and 16-bit is that used in the computer industry to signify the binary number that represents a decimal number, *e.g.* 4-bit = $2^4 = 16$, 8-bit = $2^8 = 256$ and $2^{16} = 65536$. Naturally, the greater the bit capacity the greater the complexity of functions that the microprocessor can provide and speed of reaction.

To provide the programme functions listed above a microprocessor should possess certain features:

- ROM (read only memory) which stores the programme;
- D-A (digital to analogue) convertor which converts digital numbers to analogue levels;
- I-O (input-output) interface which reads input switches and drives output solenoids, *etc*;
- RS232 (serial data line) which accepts and sends serial data from external computers;
- Timers to provide pulsing and timing facilities.

If the power source is a DC one with power transistors under feedback control from a current sensor (usually a current shunt or Hall effect device) then the microprocessor under programme control will generate the required waveform as selected by the user from thumbwheel or keyboard front panel switches, Fig. 8.1, or from external data from the RS232 serial line. On accepting a command to start welding, the microprocessor provides the waveform via the D-A convertor to the feedback controlled transistors. Further, the necessary gas flow signals and any relay commands for external peripherals can be given.

One detrimental effect may occur when HF is used to initiate the arc, because microprocessors and computers in general dislike random pulses and transient spikes as they tend to cause loss of control; also sensitive electronic components are prone to catastrophic failure in the presence of high voltages.

To avoid erratic programme control during HF initiation the circuits must be constructed with critical layout and protection methods. This usually warrants both filtering and spike protection of the relevant circuits. The programme must also possess fail safe software so that if a random spike does enter the microprocessor it does not behave erratically and cause full current output to occur, or turn on any connected peripherals.

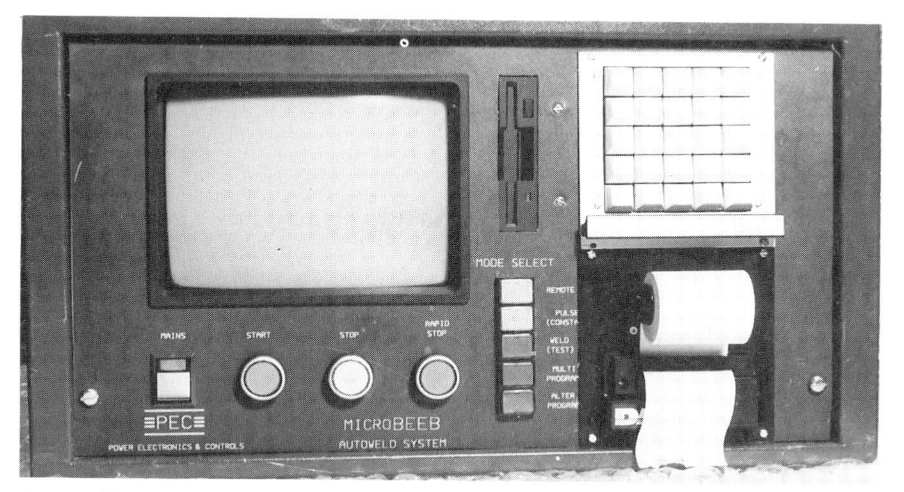

8.1 Add on microprocessor control unit for TIG welding power source with keyboard programming and hardcopy printout.

External devices

Except in the simplest of systems, automated TIG welding requires certain peripherals such as turntables, lathes, AVC, *etc*. As TIG power sources come as stand alone devices a method of interfacing all the peripherals to act in concert during welding must be realised.

A complete welding station must be able to react to various commands, *e.g.* arc established, arc delay, upslope, downslope in progress, weld completed, and others. Other commands may be added, *e.g.* gas on, pulsing operative, arc out of limits.

The easiest way to execute these commands to all the required peripherals is to sense the output commands from the power supply unit and/or the serial line to tell the individual peripheral what to do.

Should no external computer be available to read and implement the proper commands, the power supply unit must provide all the features. Hence motor controls *etc*, must be incorporated in the peripherals. This implies that the power supply unit must be rather extensive in programme capability, which is not always the case. A more common approach is to provide an external controller driven from a remote computer or, if things are simple, a programmable logic controller (PLC) which is nothing more than a rather sophisticated microprocessor with various relay, optoelectronic, voltage and current outputs.

A complex TIG welding station could probably use a PLC but to have a printed record and limit settings as well as establishing that every function is operable, an IBM PC would be the easiest to implement. It should be noted that PCs come in a variety of formats and can be just a control card, not necessarily the desktop variety. Personal computers for industrial use generally have various features such as noise suppression which all help in TIG welding.

Programming

It is no use connecting a multi-unit welding station together and expecting the system to function without proper software. Production of software or programming is a time consuming process and involves many constraints. The initial problem is the language. There are many languages currently used, the main ones are BASIC, PASCAL, C, FORTH, and ASSEMBLY.

The easiest to use is BASIC and there are various implementations. The choice depends on familiarity with programming languages bearing in mind that speed of response is of the essence.

For programming to be effective, a clear outline must exist of what the programme is to do and this is easiest to achieve by first making a list of requirements of the system under control. A typical example is:

1. Check that all units are ready;
2. Turn on gas pre-purge;
3. Initiate upslope and arc;
4. Arc established;
5. Wait for arc delay;
6. Turn on traverse, turntable, *etc*;
7. Await end of weld time;
8. Initiate downslope;
9. Await end of welding current;
10. Turn off peripherals;
11. End post-purge gas flow;
12. Ensure all units return to datum for start of new weld.

The above list is by no means comprehensive and for a practical programme to be implemented many more features would have to be incorporated, with decisions made regarding loops awaiting commands from switches, *etc*. The programme could be written from a flow diagram which consists of a format outlining all the individual requirements of the system.

Chapter 9

Exotic and difficult-to-weld metals

Many unusual metals and alloys can be welded using TIG. Amongst these are Monel metal, nimonic and other nickel based alloys, titanium, zirconium, molybdenum, *etc*, all of which can be welded with care. In the case of reactive metals which rapidly oxidise after cutting and cleaning, *e.g.* titanium, zirconium and aluminium these must be welded within a few minutes of preparing the seam. If any delay occurs the joints should be abraded or solvent cleaned again before welding.

For scraping or abrading use either stainless steel wool or a suitable proprietary dry scouring medium such as Scotchbrite. *Never* use the latter on a hot weld as it will melt and leave impurities. After abrading, wipe the edges clean with a fluff free cloth damped with solvent and allow to dry.

Preparation of reactive metals for welding is very important. With care, even beryllium copper can be welded in thin sections such as are used sometimes in production of edge welded bellows. (Beware, beryllium is highly toxic.) One point though with such metals, reruns are seldom if ever possible, so everything must be right first time. On occasions, using autogenous DC TIG, aluminium can with care be overwelded.

Titanium welds well in the DC mode and flows easily but is very prone to surface oxidation and cracking whilst cooling. For this reason the amount of purge gas needs to be increased or, when machine welding, a trailing shield can be fitted to the torch. Table 9.1 gives details of three different edge preparations for titanium butt welding, also some suggested weld parameters in the DCEN mode, gas flow, consumables, *etc.*

For critical welds in titanium it is best to operate within a chamber purged and filled with shielding gas. These chambers are commercially

Table 9.1 TIG welding conditions for butt welds in titanium, DCEN

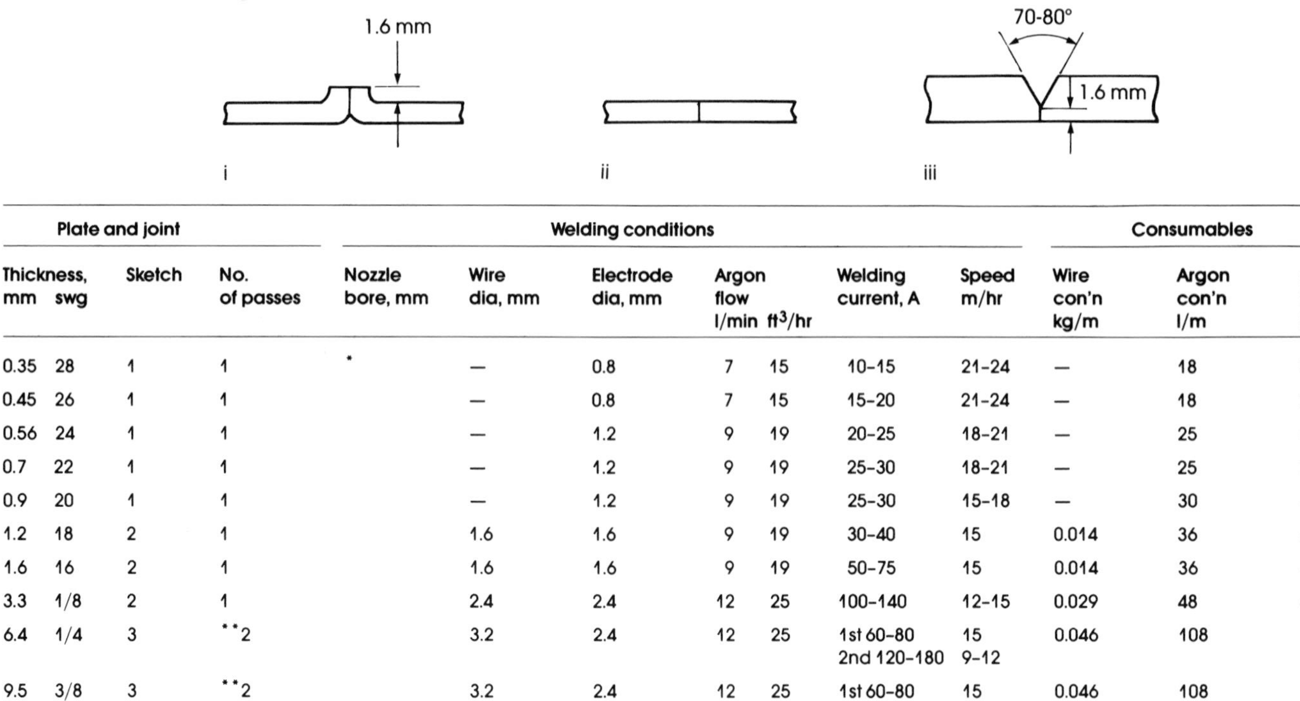

Plate and joint				Welding conditions							Consumables		
Thickness, mm	swg	Sketch	No. of passes	Nozzle bore, mm	Wire dia, mm	Electrode dia, mm	Argon flow l/min	Argon flow ft^3/hr	Welding current, A	Speed m/hr	Wire con'n kg/m	Argon con'n l/m	Arc time min/m
0.35	28	1	1	*	—	0.8	7	15	10–15	21–24	—	18	2.5
0.45	26	1	1		—	0.8	7	15	15–20	21–24	—	18	2.5
0.56	24	1	1		—	1.2	9	19	20–25	18–21	—	25	2.8
0.7	22	1	1		—	1.2	9	19	25–30	18–21	—	25	2.8
0.9	20	1	1		—	1.2	9	19	25–30	15–18	—	30	3.3
1.2	18	2	1		1.6	1.6	9	19	30–40	15	0.014	36	4.0
1.6	16	2	1		1.6	1.6	9	19	50–75	15	0.014	36	4.0
3.3	1/8	2	1		2.4	2.4	12	25	100–140	12–15	0.029	48	4.0
6.4	1/4	3	**2		3.2	2.4	12	25	1st 60–80 2nd 120–180	15 9–12	0.046	108	90
9.5	3/8	3	**2		3.2	2.4	12	25	1st 60–80 2nd 180–240	15 9–12	0.046	108	90

* *Use a large nozzle for the torch and a trailing shield for extra gas coverage.*
** *Root run: no filler.*

available and are sometimes referred to as glove compartments.

This procedure is also useful for high quality welds in zirconium and its alloys, also for aluminium if cost permits. A purge chamber does not necessarily eliminate the need for gas flow to the torch. Use a gas lens with a large bore ceramic gas cup fitted. Access to a purge chamber is by a sealed glove compartment and considerable operator skill is often required to use such equipment. Many critical welds for the nuclear industry are carried out in this way.

Welding within a purged chamber is very effective in the automatic mode. Clamping and heat sinking are of extra importance with these metals.

Phosphor bronze

Welded by TIG, phosphor bronze flows well and has a good bead appearance. As it contains copper it needs considerable heat and can seldom be welded without use of a suitable filler wire as it is extremely prone to porosity. The use of current pulsing at a rate of about two pulses per second or lower is recommended. Oscillation across the weld seam is also advantageous as it assists with sidewall fusion. This should be fairly slow, perhaps one cycle per second, easy to achieve by mechanical means. Most copper-nickel alloys benefit from using this technique. Check carefully that the filler rod or wire is compatible with the base metal.

Free machining steels

Mention must be made here of these very difficult-to-weld metals. The addition of higher proportions of sulphur and graphite can give rise to extreme porosity and they should be avoided for welding whenever possible. If welds are essential, improvements can be made by careful use of heat sinks and arc pulsing as the pulses often bring inclusions to the bead surface. The real problems arise in autogenous welding but here the addition of suitable filler wire can, as with many other metals, reduce porosity considerably.

Stainless alloys

Stainless steels can have cast-to-cast variations in alloy content which, however slight, may produce unacceptable welds. For critical aerospace, nuclear and process welds obtain metals with a certificate guaranteeing the batch quality and alloy percentages. Most stainless steels such as types 308, 316 and 347 are considered very weldable for most purposes. For special alloys, experiment carefully first. Always consult a manufacturer of filler wires for compatible materials as a vast range of wires is available.

Brass

Some types of brass can be welded with DC TIG but the generally large zinc content makes these metals very difficult to join. The author's advice is not to bother. Other joining methods, *e.g.* brazing are much less trouble and more cost effective.

Aluminium welding

AC mode

It is generally considered in the welding industry that aluminium, magnesium and their alloys should always be welded by the AC TIG process. It is true to say that the AC mode is more successful because the alternating current has some cleaning effect on the weld bead, bringing unwanted oxides to the surface, Fig. 10.1.

This mode is particularly successful for thicker metals using filler wire and is almost mandatory where aluminium filler wire or rod are being used. One slight disadvantage is that AC TIG can sometimes produce considerable sparking and spatter but even this can be minimised with proper cleaning of the weld preparation. Aluminium should be welded immediately after cleaning the weld seam surfaces for best results. Keep the weld area free from any contaminants which might be a potential source of hydrogen porosity.

DC mode

Autogenous welding of some aluminium alloys can be carried out by DCEN but it is usually only possible with fairly thin sections and with pulsing of the arc. It should be noted that aluminium welding often involves considerable experimentation before good results are achieved, because two

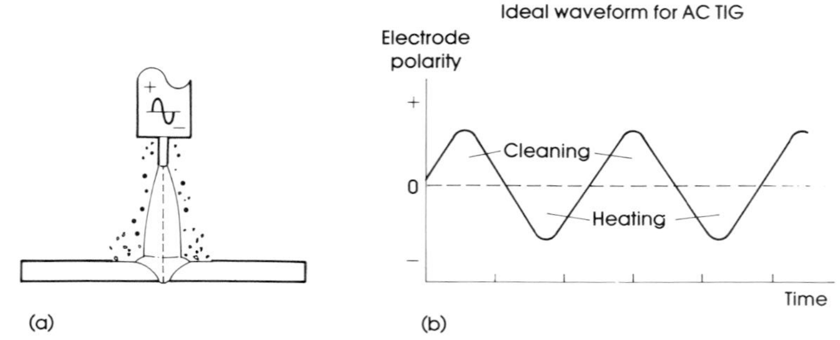

10.1 AC TIG welding arc effect: a) Good cleaning action and weld preparation; b) Ideal current waveform.

batches of what is ostensibly the same alloy often differ considerably in weldability. Generally speaking the purer the metal the better it will weld by DC. The following have given successful results:

- Pure aluminium;
- N3, N4, N8, H30;
- 2219, 5083 (N8), 5454, 6082 (H30). Avoid *all* 2000 series alloys.

CONDITION

Fully annealed is best but half hard can be welded with care.

CLEANNESS

Lightly abrade, solvent clean and weld within ten minutes.

JOINTS

A tight, square, closed butt preparation is best. Back purge the underside of the seam if possible. Ensure an even heat balance either side of the seam and clamp faces firmly to avoid distortion.

ARC GAP

This is critical. Either make *very* accurate piece parts or use arc length control (ALC) to maintain a constant gap.

WELD RUN

Use *minimum* overlap at end of weld. A second run over a seam often spoils the weld. Too long a downslope time has the same effect as too much overlap.

METAL THICKNESS

For all practical purposes maximum thickness is 2 mm for an autogenous butt weld but conditions vary from alloy to alloy.

SHIELDING GASES

Argon gives good results on most occasions. Helium is more satisfactory but expensive. Satisfactory gas spread and arc strike are difficult to achieve in helium, which can also totally inhibit pulsing of the arc.

WELDING CURRENT – PULSED DC

High, short, primary pulses and low, long, background pulses give best results, Fig. 10.2.

Typical parameters

Primary current:	64 A	Background current:	8 A
Primary pulse:	30 msec	Background pulse:	150 msec
Upslope:	Nil	Downslope:	2 sec max

Keep higher or lower currents within the same average ratios.

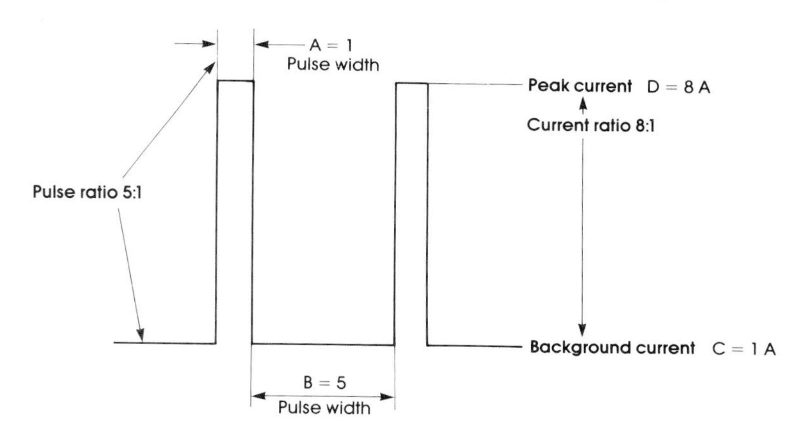

10.2 Recommended waveform for pulsed DC TIG welding of aluminium and its alloys.

WELDING SPEED

Determined by experiment. Speeds up to 16 mm/sec (36 in/min) have been achieved with thin materials but some alloys need to be welded much more slowly. Avoid remelt runs.

CLAMPING

Must be firm. Any heat sinking and watercooling requirements must be determined by experiment.

HELIUM

This increases heat input into the weld and can give a better cosmetic finish. However, arc strike is difficult in helium and it may be necessary to establish an arc in argon and then switch over as quickly as possible to helium. In some circumstances, particularly when using helium in a pressurised chamber, pulsing the arc may not be possible. Try to use low (around 2 bar) gas pressure with good flow and spread. A gas lens collet assembly in the torch is advantageous too and will maximise localised gas coverage within the chamber.

ARC STRIKE

Set any strike current level control on the power source to *maximum* (usually around 20 A) when using helium.

Aluminium has extremely good thermal conductivity and thus needs a higher current at all times than steels. Try to concentrate the arc as much as possible by keeping the arc length short and consistent.

ELECTRODES

When used for AC TIG welding electrodes have a comparatively short life compared to DC use and are generally made from pure or 0.8% zirconiated tungsten. Figure 10.3 shows a preferred method of tip grinding. During welding a smooth surfaced ball forms on the end and this gives the correct arc form. For DC TIG welding a sharp ground point can be used.

EDGE PREPARATION

Table 10.1 gives details of four different edge preparations for aluminium butt welding, together with some suggested welding parameters, gas flow, consumables, *etc*. Further parameters are given in Table 10.2 (with Table 10.2 refer to chapter 6 for reference to the listed joint types).

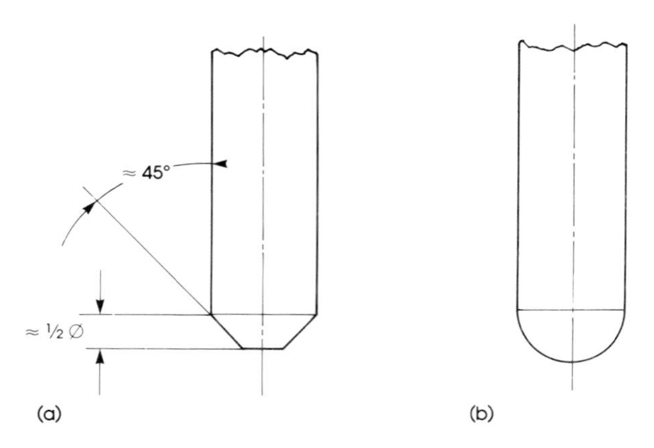

10.3 Electrode tip shape for AC TIG welding: a) As ground; b) During welding.

Table 10.1 Manual TIG welding conditions for butt welds in pure aluminium, AC, flat position

Plate and joint				Welding conditions						Consumables		
Thickness, mm	swg	Sketch	No. of passes	Nozzle bore, mm	Wire dia, mm	Electrode dia, mm	Argon flow l/min	ft^3/hr	Welding current, A	Speed m/hr	Wire con'n kg/m	Argon con'n l/m	Arc time min/m
0.9	20	1 or 2	1	9.5	1.6	1.6	5	10	45–60	21	0.007	14	2.8
1.2	18	2	1	9.5	2.4	2.4	5	10	60–70	18	0.018	17	3.3
1.6	16	2	1	9.5	2.4	3.2	5	10	75–90	18	0.024	17	3.3
2.0	14	2	1	12.7	2.4	3.2	5	10	90–110	18	0.028	17	3.3
2.6	12	2	1	12.7	3.2	3.2	6	13	110–120	18	0.034	20	3.3
3.3	10	2	1	12.7	3.2	3.2	6	13	130–150	17	0.047	21	3.5
4.8	3/16	3	1	12.7	3.2	4.8	7	15	150–200	15	0.09	28	4.0
6.4	1/4	3	1	16	4.8	4.8	7	15	200–250	15	0.13	28	4.0
9.5	3/8	3	2	16	4.8	6.4	8	17	270–320	10–12	0.22	87	10.9
12.7	1/2	4	2	16	6.4	8.0	9	19	320–380	9–10	0.28	108	12.0

Zirconiated electrodes are preferred

Table 10.2 Further welding conditions for AC welding of aluminium

Metal thickness, mm (in)	Joint type	Tungsten electrode diameter, mm (in)	Filler rod diameter mm (in)	Welding current, A	Argon flow, l/min (ft^3/hr)
1.5 (1/16)	Butt Lap Corner Fillet	1.5 (1/16)	1.5 (1/16)	60–85 70–90 60–85 75–100	7 (15)
3.0 (1/8)	Butt Lap Corner Fillet	2.0- 3.0 (3/32–1/8)	2.0 (3/32)	125–150 130–160 120–140 130–160	7 (15)
4.5 (3/16)	Butt Lap Corner Fillet	3.0-3.8 (1/8–5/32)	3.0 (1/8)	180–225 190–240 180–225 190–240	10 (20)
6.0 (¼)	Butt Lap Corner Fillet	3.8–4.5 (5/32–3/16)	4.5 (3/16)	240–280 250–320 240–280 250–320	13 (25)

Chapter 11

Design of weld joints for TIG

Figure 11.1 shows a fair selection of the more common seam and edge preparations for TIG welding. The author does not recommend that *any* edges which are to be TIG welded be laser cut. On stainless metals in particular, laser cutting often leaves a very slight burning or charring of the edge, not always visible to the naked eye but which can lead to inclusions and porosity in the weld. Guillotined or machined edges are always preferable and must be as clean and true as possible. Try to achieve a nil gap condition at all times. Some comments on the various forms in Fig. 11.1 now follow:

A CLOSED SQUARE BUTT PREPARATION – HAND OR MACHINE WELDING

This is the most common, particularly when joining thin sheets and thin wall tubes. It is generally agreed that 2.5 – 3.0 mm (0.100 – 0.120 in) is the maximum thickness which, dependent on material, can be autogenously welded using this preparation.

B, C, D AND E MACHINED PREPARATIONS FOR HEAVY PLATE OR PIPE WALL – HAND OR MACHINE WELDING

These preparations can also be used in the *open root* condition. The best preparation for a specific weld can be found from experience or experiment. Preparations D and E probably give better sidewall fusion but are most costly to machine. See also note 1 below.

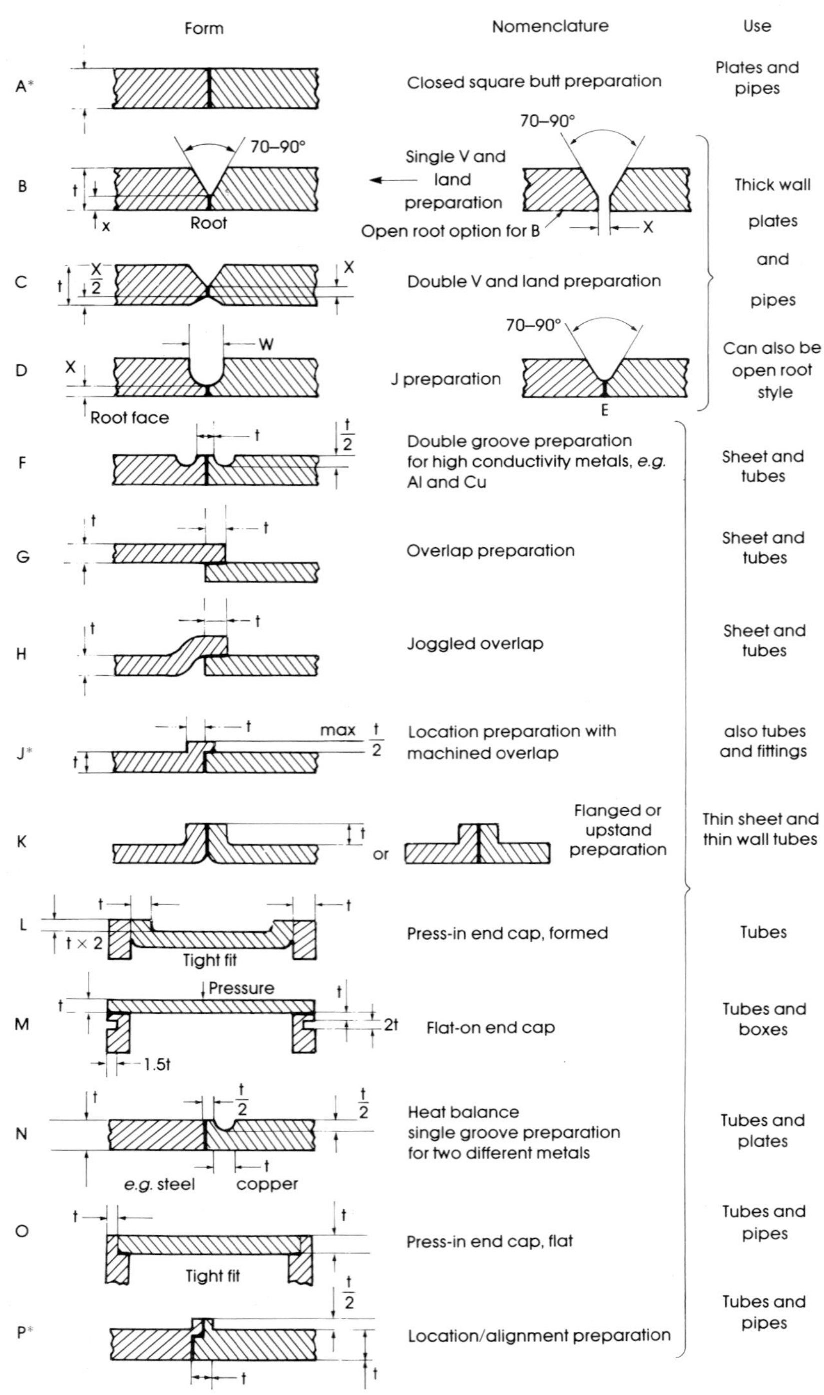

*Also for internal bore welding, t max 2.5 mm.
t = total metal thickness, X = face width at root, W = width of J preparation at top surface.
For preps B, C and D the value of X seldom needs to exceed $\frac{t}{4}$, using an autogenous weld run.

11.1 Selection of joint preparations for TIG welding.

F DOUBLE GROOVE PREPARATION – HAND OR MACHINE WELDING

This preparation is useful for copper and alloys with high thermal conductivity. It has a slight disadvantage in as much as it leaves a depression (undercut) in the top surface of the weld bead with subsequent reduction in strength.

G OVERLAP PREPARATION – HAND OR MACHINE WELDING

A good preparation for alignment of tubes but difficult to weld unless some really sophisticated heat sinking is applied. The material either side of the overlap could melt back if the electrode is not correctly positioned and incorrect welding parameters used. Best used with arc pulsing.

H JOGGLED OVERLAP PREPARATION – HAND OR MACHINE WELDING

Similar to G with the same problems. Should not be used for welds which require top quality mechanical strength.

J LOCATION PREPARATION WITH MACHINED OVERLAP – MACHINE WELDING

Excellent for alignment, this preparation puts a small amount of extra metal into the weld pool. Can be welded externally or internally for thick walls. Widely used for welding end fittings to small bore tubes in the process industries.

K FLANGED OR UPSTAND PREPARATION – HAND OR MACHINE WELDING

Ideal for thin aluminium sheet.

L INSERTED END CAP FOR TUBES AND CANISTERS – HAND OR MACHINE WELDING

The end cap should be a tight fit and held in by a heatsink to prevent tipping and distortion when welding. Very suitable for items such as aluminium battery cans.

M FLAT-ON END CAP FOR TUBES AND CANISTERS – MACHINE WELDING

This can be used for welding the sensor plate to a transducer body, a bursting disc to a liquid pressure overload device, pressure plates to load cells, *etc.* Align the disc carefully and hold in place with a suitable heat sink. Also suitable for some battery cans.

N HEAT BALANCE PREPARATION – HAND OR MACHINE WELDING

A single groove is cut into the metal which needs the most heat. The easier melting metal needs careful heat sinking close to the weld line. Not all metals are compatible so check first whether the two metals *will* mix and weld.

O PRESS-IN END CAP FOR TUBES AND CANISTERS – MACHINE WELDING

Similar to L. Both these preparations are useful for welding a thin disc to a thick walled tube.

P LOCATION PREPARATION, PIPES AND TUBES – MACHINE WELDING

This preparation adds some extra metal to the weld pool and can be used for either external or internal welds. Should be made with a tight fit. Similar to J.

NOTES

1 Root runs for preparations with open gaps such as B, C, D and E should be kept as small as possible consistent with strength. Distance x, *i.e.* the face width at the root, should be roughly proportional to the thickness t in about a 1:4 ratio but is seldom less than 1.5 mm in practice;
2 Any preparation where one side is located within the other such as J, L, O and P should always fit tightly and have clean faces;
3 Proportions given in Fig. 11.1 are by no means mandatory but given as a guide;
4 Preparations G and H need great care with arc positioning;
5 Try to machine preparations with a minimum of cutting oil or fluid, and clean off any surplus oil with a solvent before welding. Ground edges are not recommended because of possible grit inclusions.

Open gap preparations for root runs

These preparations, all variations of types B, C, D and E in Fig. 11.1, are very useful for obtaining a full bead with good sidewall fusion.

Ensuring that the bead does not droop when carrying out an open gap run demands great skill on the part of the operator. This closing run is critical to weld viability and it is here that pulsing the arc can be of great help by allowing the bead to freeze between bursts of high current. This applies to both autogenous and filler wire runs. After the root run has been completed the object is to fill the gap with metal in the strongest and most economical manner, at the same time achieving maximum penetration into the side-walls.

Consumable weld seam inserts

Mention of possible droopthrough when carrying out root runs in open gap welding leads naturally on to consumable inserts. These take the place of filler rod or wire and are available in the form of either straight extruded strips for longitudinal seams or, more commonly as rolled split rings to be inserted when welding tubes where they come in a wide variety of sizes, forms and materials. Shown are some of the most popular forms, Fig. 11.2, although some suppliers will produce specials in small quantities.

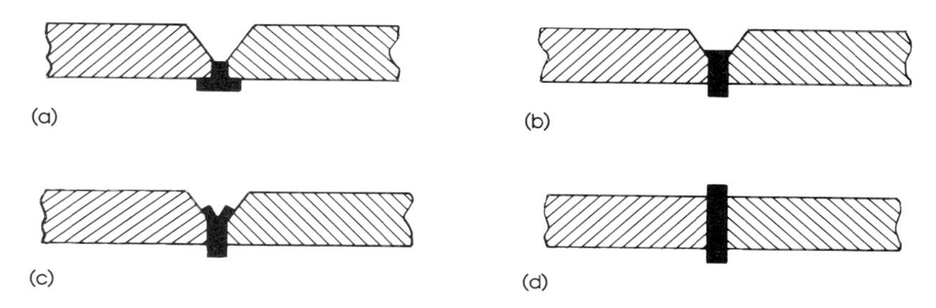

11.2 Common types of consumable insert: a) EB insert; b) Tacking not needed; c) Tacking not needed; d) Needs tacking.

Some shapes such as (b) and (c) stay in place without tacking but others need this. Always ensure that inserts are made from compatible metals and have sufficient but not too much metal bulk on melt down. They are useful for filling root runs in open gap V or J preparations, especially for mechanised orbital welding as they often obviate the need for a wire feeder. Attention must be paid to the following:

- Arrange for some slight lateral pressure to be put on the component parts either side of the joint when welding starts. This pressure is only to hold the gap, not to close it or the bead will be squeezed out;
- Track the torch accurately along the seam centreline;
- Use arc pulsing wherever possible to freeze the bead between pulses. This prevents droopthrough;
- Always back purge if convenient but keep internal build-up of gas pressure to a minimum. Sometimes nil or even negative pressure is an advantage to assist with full underbead formation.

Wire feed for open gap – for filler runs on V and J preparations

Use a feeder which gives pulsing of the wire in synchronisation with the level and time of background current. This facility is generally found only on the most expensive power sources and feeders.

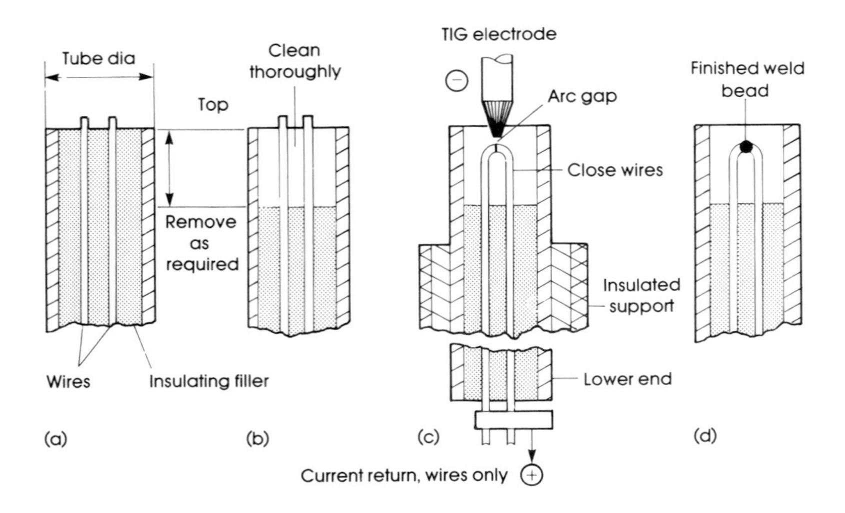

11.3 Thermocouple junction welding: a) As cut; b) Ends prepared; c) Ready to weld; d) Weld completed.

Thermocouple tip welding

Figure 11.3 shows preparations for TIG welding the two internal wires at a thermocouple tip.

Points to remember here are:

- Total cleanness of the area where welding takes place is needed;
- Accurate electrode-to-work gap and tip positioning are essential;
- Suitable heat sinking must be provided;
- Very precise arc-on timing and consistent current values have to be maintained;
- Good shielding gas coverage must be present.

Thermocouple wire welding is a very exact technology and many ingenious solutions have been found.

Tube to tube plate (or tube sheet) weld preparations

Equipment to carry out this task can be very expensive and is briefly discussed in chapter 7. Some basic weld preparations for external welds are shown in Fig. 11.4. Try to design joints for autogenous welding as it is difficult and even more expensive to use equipment with built-in wire feed to the arc area. Ensure that preparations are machined to give the tube a good fit

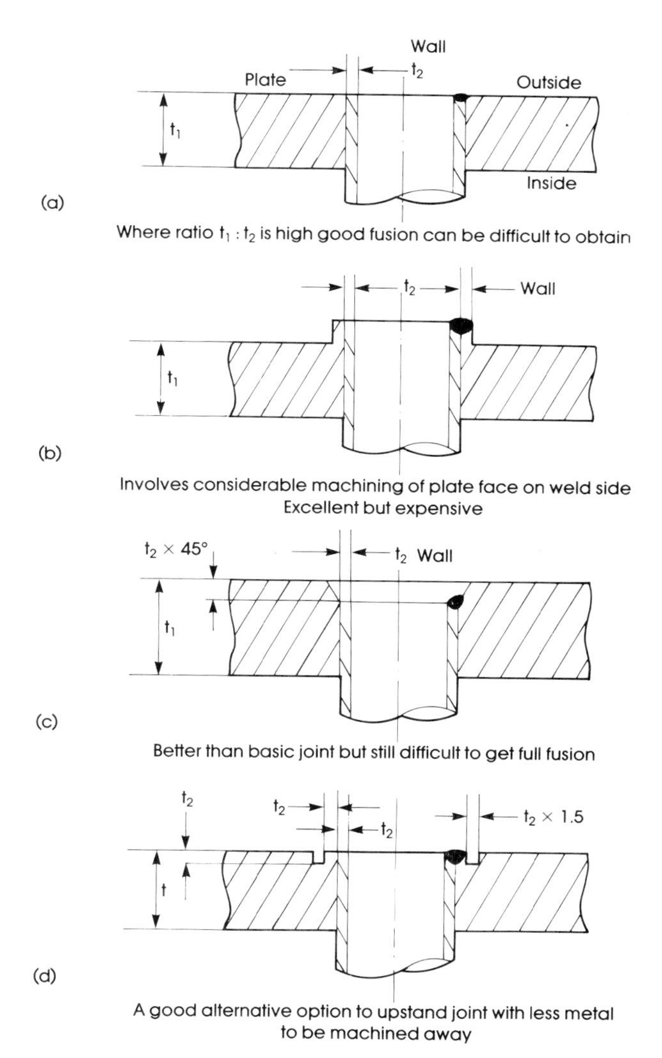

11.4 External tube to tube plate weld preparations: a) Basic joint; b) Upstand joint; c) Countersunk basic joint; d) Recessed upstand joint.

in the tube plate. Mechanical expanding machines both pneumatic and hydraulic are commercially available and of great assistance in achieving such fits.

Internal welds can be made to ensure that the outer face of the tube plate remains clean and free from weld spatter. Special internal orbital welding heads are needed for this process and it is especially advantageous to make all such welds autogenously as the tubes are of small bore on many occasions, Fig. 11.5.

Internal welds often need slight longitudinal pressure to be applied to the joint whilst welding takes place.

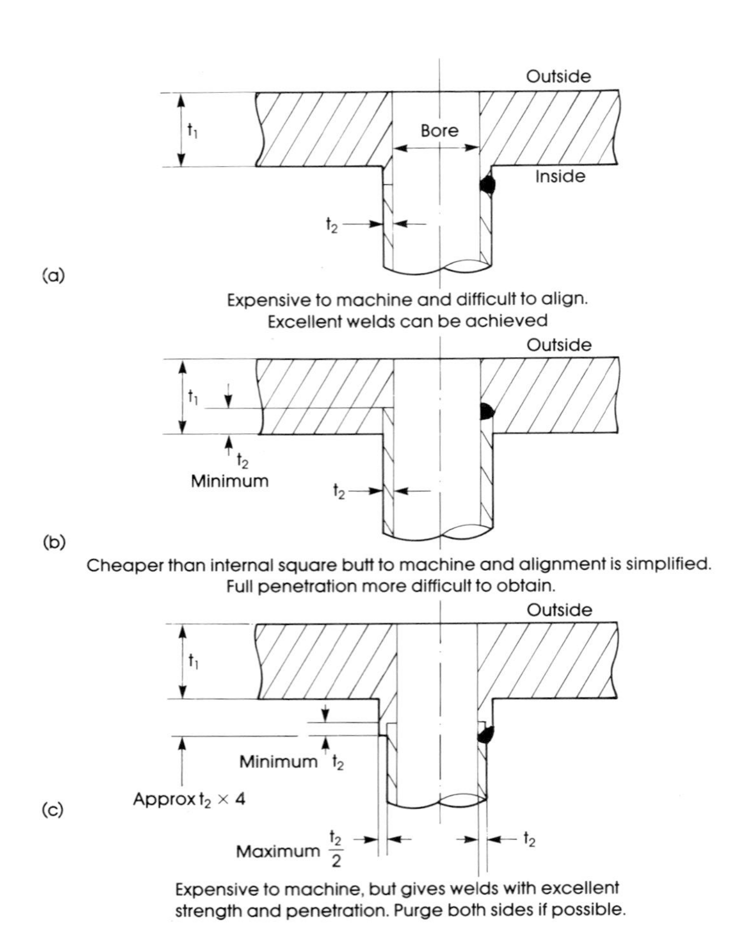

11.5 Internal tube to tube plate weld preparations: a) Square butt; b) Self-aligning; c) Spigot alignment.

Backing bars and rings

To ensure a good underbead when carrying out weld seam root and capping runs, particularly for an open gap root run, it sometimes becomes necessary to provide a removable backing bar or ring which supports the molten underbead, preventing droopthrough and also provides heat sinking either side of the weld seam. Figure 11.6 shows some configurations and chapter 12 lists suitable materials and their advantages and disadvantages.

Internal backing bars inside pipes and tubes should always be made split or segmented for easy removal when welding is completed. Figure 12.2 gives some groove proportions.

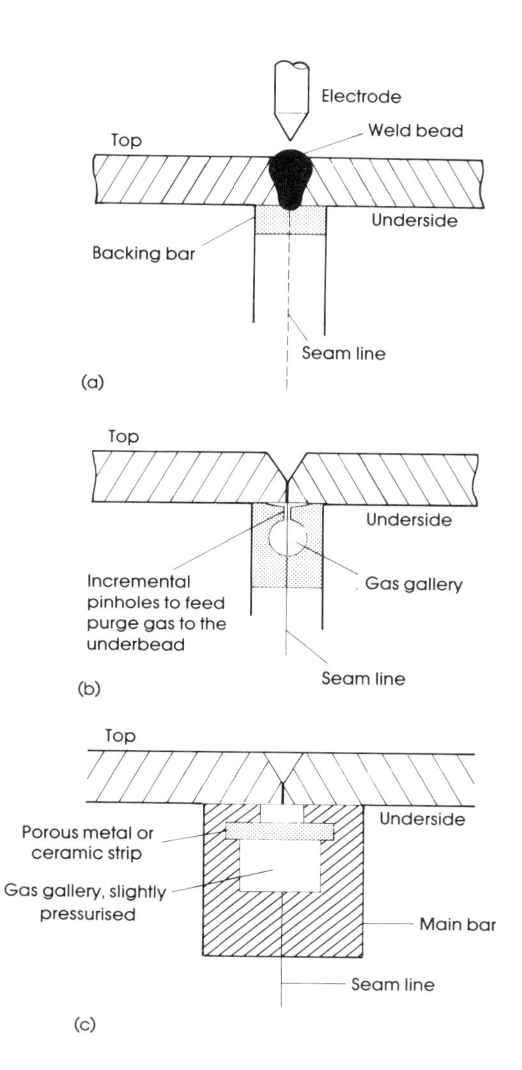

11.6 Backing to: a) Support weld underbead; b) Purge and support underbead; c) Provide back purge only.

Chapter 12

Jigs, fixtures and heatsinks for automated TIG welding

Most of the skill required to carry out machine welding successfully is in design of the fixtures that hold the components together whilst welding takes place. In fact the author considers that the mechanical items account for about 85% of problems. Power sources are very accurate and by using the correct electrodes and gases produce almost any type of arc required.

Welding engineers contemplating automation must develop a feel for a particular job and can soon attain the necessary know-how to design and produce suitable tooling, the aim being to transfer unwanted heat build-up away from the weld area, at the same time holding the joint faces together to produce viable welds with maximum penetration and strength. Chapter 7 has shown some of the machinery required: this chapter gives advice on tooling and clamping.

Heatsinks or chills

These can be made to provide clamping or merely laid in place alongside the seam. It is most economical if they can carry out both functions. The first problem is to find a suitable material, this must be chosen for its heat transfer properties, strength and life and, by its nature, have no effect on the parent metal when heated. Some metals outgas when heated which produces contamination, perhaps only slightly, in many critical joints. Materials may be as follows:

- *Copper*. This may require plating to ensure that no copper contamination takes place from rubbing. Excellent heat transfer properties;
- *Stainless steels*. These have a low melting temperature and are best used for fine, low current pulsed welds, *e.g.* edge welded bellows;
- *Mild and carbon steels*. Not recommended but satisfactory for non-critical welds. Easy to machine and cheap to replace. Prone to rust;
- *Brass*. Not recommended. Easily machined and cheaper than copper but a source of possible contamination. Could also be plated.
- *Ceramics*. Specially made for a particular job the use of these has not yet been fully investigated. Consult a ceramics expert first;
- *Aluminium*. Like copper, aluminium has good heat transfer properties but is soft and easily damaged. Use for smooth components which will not easily damage the heatsink.

The above materials have been suggested for manufacture of heatsinks and backing bars. Two typical precision heatsink types and the proportions involved are shown in Fig. 12.1.

Precision welding

The word precision implies that finished components are accurate and clean with minimum distortion, another factor in heatsink design. They do not always need to be in tight contact with the component as this sometimes makes it difficult to remove a finished item after welding, a point often overlooked. It is good practice to include some form of ejection if possible, which can save burnt fingers. Remember that a heatsink stores heat, so give the fit some allowance for thermal expansion.

Clamping jigs for machine seam welding

The principles involved here are shown in Fig. 12.2 together with suggested proportions. The clamp toes are usually made from steel or copper, or steel tipped with copper to save replacement expense.

Segmented or finger clamps are useful to ensure even pressure along the seam length. The layout shown is that which would be used for thin sheet welding with a square closed butt edge preparation. Other suitable edge preparations are shown in chapter 11.

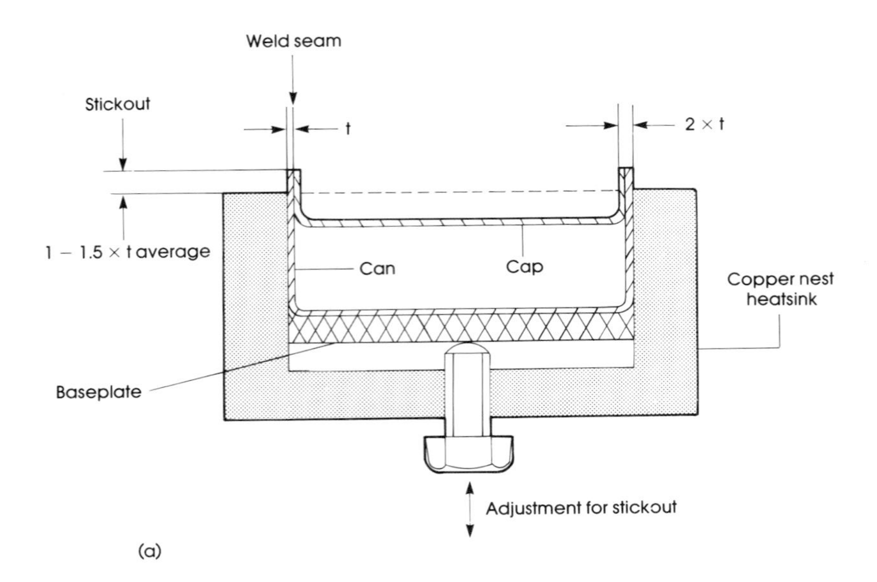

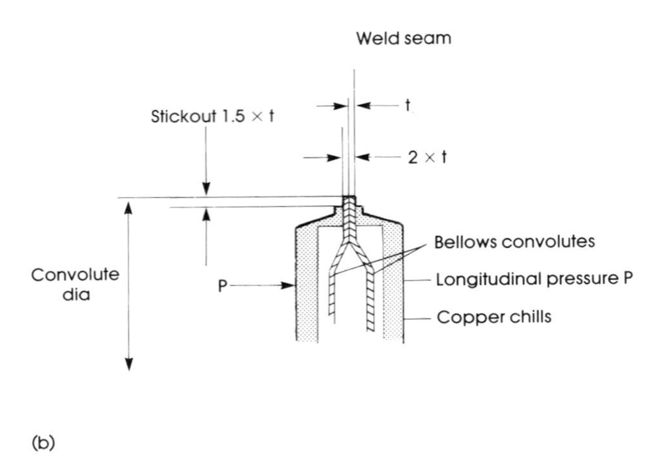

12.1 Special heatsinks: a) Nest type for battery canister seals; b) Copper chills for edge welding of bellows.

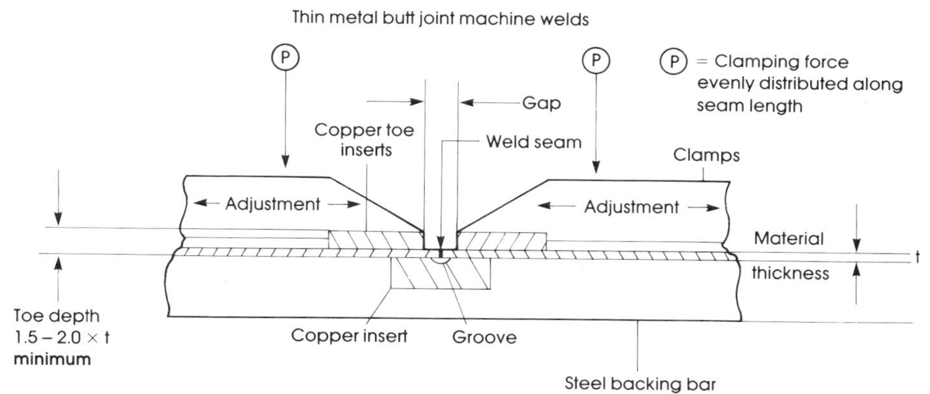

12.2 Clamping jig for mechanised longitudinal seam welding.

Notes on Fig. 12.2

- Gap for metal thickness 0.08–0.5 mm (0.003–0.020 in) = 0.8–2.0 mm (0.030–0.080 in).
- Gap for metal thickness 0.5–2.0 mm (0.020–0.080 in) = 2.0–4.0 mm (0.080–0.160 in).
- Gap for metal thickness over 2.0mm (0.080 in) = 1.5–2.0 times t.
- Groove: width 2 × t; depth 1 × t or 0.25 mm (0.010 in), whichever is smaller.

  Metals less than 0.25 mm (0.010 in) thickness may need no groove in backing bar.

  The gap should be evenly astride the weld seam.
- Clamps should be independently adjustable.
- Clamps and backing bars should be either made completely from copper, or steel with copper toe inserts.
- For even clamping over the seam length use segmented fingers, particularly on long seams.
- Good values of P per unit length of weld are:
  - Metal up to 1.0 mm (0.040 in) thickness: 10 N/mm (50 lbf/in) per side;
  - Metal over 1.0 mm (0.040 in) and up to 3.0 mm (0.120 in) thickness: 20 N/mm (100 lbf/in) per side;
  - Over 3.0 mm (0.120 in) much higher forces may be needed.

Remember, firm accurate clamping is a must. Values given are for guidance only.

Hand welding clamps

Although use of clamps is most advantageous in machine welding, fairly simple devices can free the operator's hands as well as allowing him more room for error in heat input. Hand operated toggle clamps or similar can be used, being comparatively cheap and easily replaceable. Magnetic clamps are *not* recommended for use with manual TIG, as they have a tendency to deflect the arc if placed too near the seam.

Clamps reduce HAZ size and also make for a reduction in final cleaning and polishing of the finish welded joint, particularly in the case of machine welded metal cabinet top corners, Fig. 12.3. Tops welded thus require only degreasing before going for spray or powder painting.

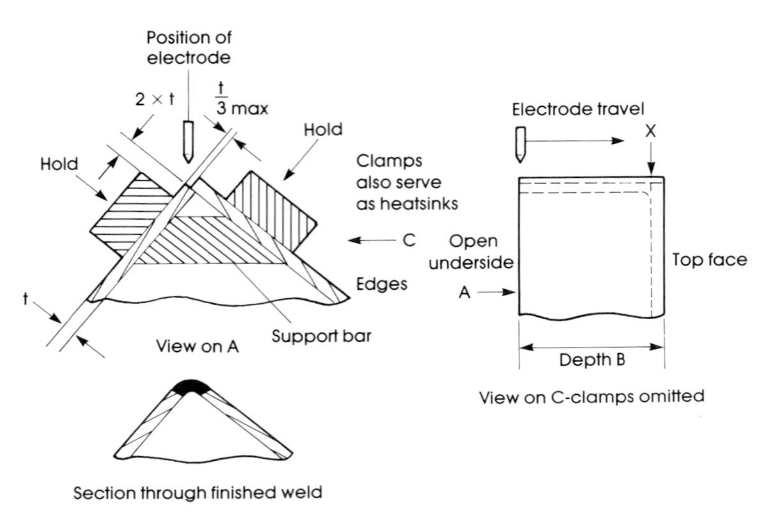

12.3 Suitable joint configuration and clamping for machine welding sheet metal cabinet top corners.

Notes on Fig. 12.3

The proportions given are approximate only. Pulsed welding is preferable and the metal should be wiped before welding but not necessarily solvent cleaned as the TIG arc can cope with a very thin film of oil on the metal surfaces. The gap along the weld seam must be completely closed.

The weld sequence should be as follows:

1 Close clamps – purge gas on – arc strike;
2 Short upslope and electrode travel dwell;
3 At full welding current electrode travel starts;
4 Short downslope should commence at point X;
5 Extinguish arc when electrode is aligned with top face;
6 Purge gas off. Retract clamps, remove finished component.

A cabinet top where dimension B = 30 mm can be welded in 7–10 seconds using this method. A fully specified power source is required.

Summary of tooling, fit-up, heat sinks and clamping

Attention to these points is every bit as important as care taken over the actual welding process. An automatic weld is only as good as its tooling. So check out the following:

- Heat balance when welding dissimilar materials, *i.e.* metals with differing thermal conductivity, is *critical*;
- Try also to achieve optimum heat balance when welding thick to thin materials;
- For high speed welding try for maximum heat isolation;
- Try to allow for extra component material to be available for a good melt down. This can be part of the original material when performing autogenous welds, filler wire or rings for thicker materials;
- Keep all joints clean and contamination free before welding and keep seam edges sharp. Use guillotined rather than laser cut edges;
- Heatsinks or chills improve heat balance and can in addition sometimes be used for clamping;
- Make all clamping and manipulation devices as accurate as possible, with smooth progressive motion;
- Ensure first class current return (earthing).

Some important alignment details are given in Fig. 12.4 and its accompanying notes.

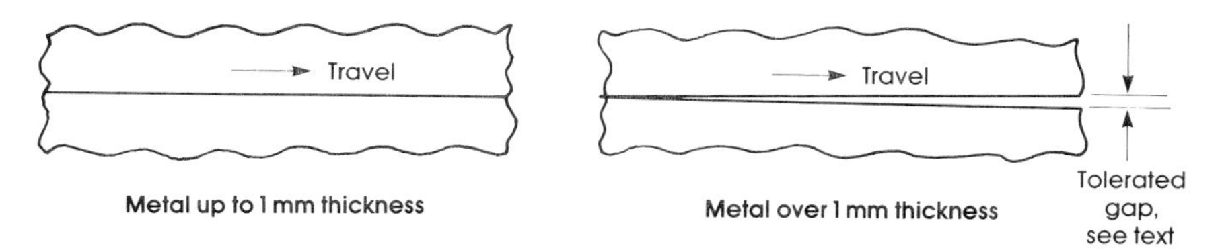

12.4 Alignment of longitudinal weld seams.

Notes on Fig. 12.4

- All butted edges must be straight, clean and square cut with no burrs;
- Tolerated gap in (b) should always be less than 10% of the material thickness. This tolerated gap applies only to metal over 1 mm (0.040 in) thickness, where progress of the arc along the seam often pulls the edges together. Many companies making rolled seam welded tubes use this technique successfully;
- For very thin metal observe almost clinical cleanness and use a good non-inflammable solvent to remove excess oil and grease. Use a lint free cloth;
- Take off any slight burrs with a fine dry stone lightly applied to the edges, then wipe and solvent clean again.

Chapter 13

Ancillary equipment

This chapter is concerned with equipment and devices which add to the efficiency of the TIG process. No item of equipment will be described at great length or in detail, neither will any specific recommendations be made. Many specialist companies produce equipment as described. Obtain information from them and a demonstration wherever possible. Some equipment is tried and tested, some is new and still in the development stage.

Wire feeders

When compared with feeders used for MIG/MAG welding, TIG wire feeders need to be much more comprehensively equipped. In addition to a lower overall wire speed range it is usual to provide all or some of these features:

- Wire speed range down to 0.5 mm/sec and up to around 18 mm/sec;
- Instant stop/start of wire by remote control;
- A burnback facility at switch off to retract the wire from the weld pool and prevent it freezing in. Usually this switch off and retraction occurs just before commencement of downslope but this time should be variable to suit;
- Variable pulsing of the wire feed synchronised with arc pulsing;
- Accurate and consistent wire speed;
- Instant switch off if the electrode stubs in to the pool;

- Four roll wire drive mechanism for smooth operation and to give some wire straightening.

Manufacturers of TIG wire feeders offer a wide range of optional features. This makes them rather expensive but for critical welds in expensive components the cost is justified.

At the point where the arc occurs and the feeder nozzle is situated, several adjustment facilities are needed, Fig. 13.1.

Wire nozzles should be made from a good grade of hard copper and easily replaceable. Keep distance A as short as possible to avoid any tendency for the wire to curl. On automatic TIG set-ups the wire is often supplied tightly wound on to special 100 mm (4 in) diameter spools and needs straightening before it emerges from the nozzle tip. Four roll wire feed drive units solve this problem to a great extent.

A TIG wire drive unit should be able to accept wires from 0.4 mm diameter (not a standard size) up to around 1.6–2.0 mm diameter.

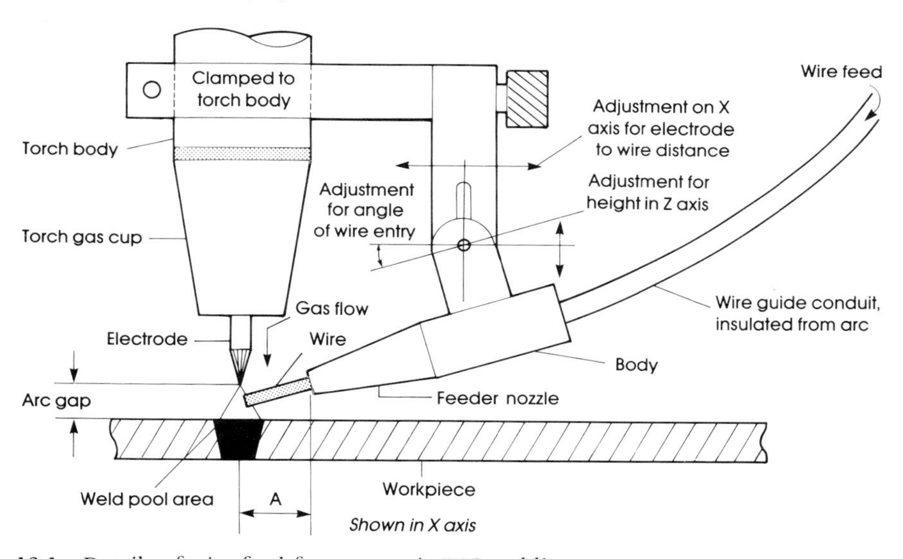

13.1 Details of wire feed for automatic TIG welding.

HAND WELDING WIRE FEED

An ingenious wire feed system is available with the wire fed directly to the torch, Fig. 13.2. This gives the operator a free hand and is most useful in confined spaces.

Arc oscillation – methods and equipment

As has already been mentioned in previous chapters it is advantageous when seam welding thick sections using a V or J preparation to oscillate or

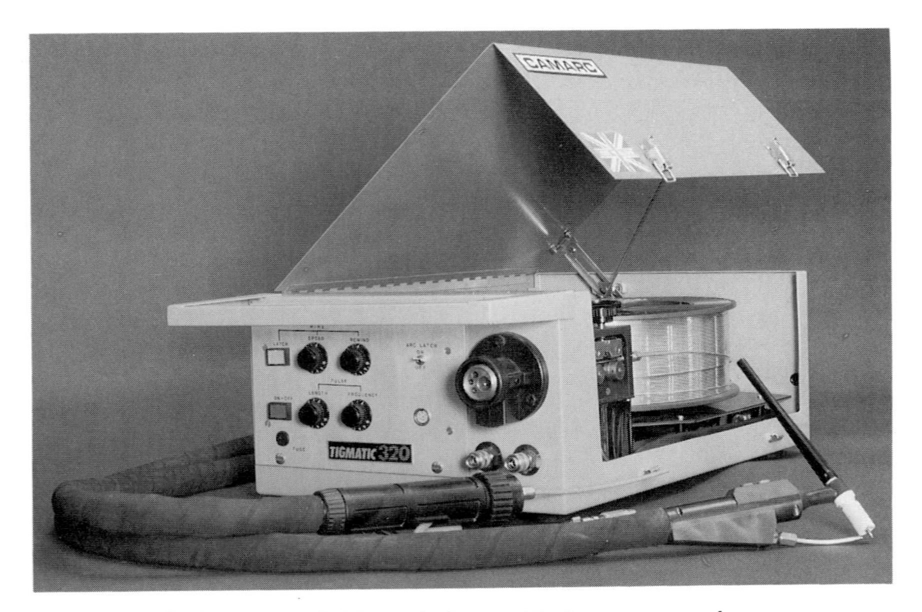

13.2 Wire feed system with 'through-the-torch' placement nozzle.

weave the arc laterally across the seam. This assists in obtaining optimum edge melt in or sidewall fusion. Usually employed in automatic welding this oscillation can be carried out in two ways. One is by mechanically moving the torch nozzle with the electrode in its gas cup across the seam at a predetermined distance to travel rate ratio. The other is by magnetically bending the arc, using double or multiple magnet poles positioned either side of the arc and switching polarity to deflect the arc to one side or the other in sequence as required. Arc switching can in both cases be synchronised with arc pulsing. Both systems offer a good solution to the problem of successfully carrying out large TIG filler runs, in particular.

Seam followers - for mechanical systems only

With a straight seam and a large V or J preparation, there is little difficulty in following the seam by guiding the electrode down the centreline using a mechanical follower probe travelling a short distance ahead of the arc. This probe when moved laterally sends signals back to servo-motors which correct the arc position in relation to the seam deviation which it has sensed. This type of mechanism is rugged and can be used either with longitudinal seams for flat plates or circumferential seams in pipes and tubes. However, seams in the latter are usually machined true with a suitable preparation before welding takes place so a follower is needed only if seam accuracy is in doubt.

A probe needs faces or an edge for guidance so grooved seam preparations are ideal. However, for an almost invisible seam, *e.g.* a closed square butt, something more sophisticated is needed, perhaps an infrared beam, a magnetic sensor, a laser beam or an acoustic probe. These all have an advantage in that they are non-contacting devices. To summarise, check which system suits your needs and purchase equipment to suit from a reliable supplier. Many seam followers are very expensive and dedicated to one job only.

Internal gas purging systems

When working on welds in, say, large diameter pipes where gas back purging of the weld underbead is essential, it is unsatisfactory and uneconomical just to flood the inside of the pipes with shielding gas. There are several methods of achieving economy with efficiency in this area.

PURGE DAMS

A paper or card dam can be arranged in the tube, Fig. 13.3. The disadvantage of this method in a closed or inaccessible system is removal of the dams after welding is completed. A proprietary range of semi-stiff soluble paper circles is available for making dams and these are dissolved and flushed away when no longer needed using large quantities of water.

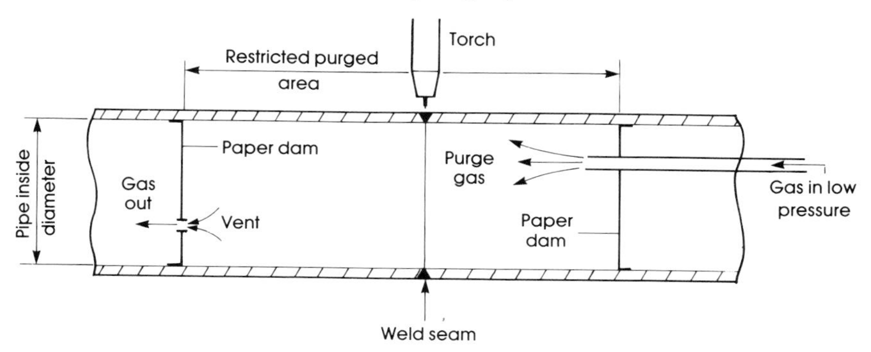

13.3 Paper dams for back purging.

BLADDERS

A much superior and reusable method of back purging is by a system of inflatable dams, more usually known as bladders, Fig. 13.4. The bladders are inflated by air or, for small diameters, by the shielding gas itself. After use the gas is cut off, deflating the bladders and allowing the system to be pulled out of the pipe by a chain or cord. When using such a system *never* allow too much gas pressure to build up within the purging region as this will produce deformed underbeads or in extreme pressurised cases weld pool blowout.

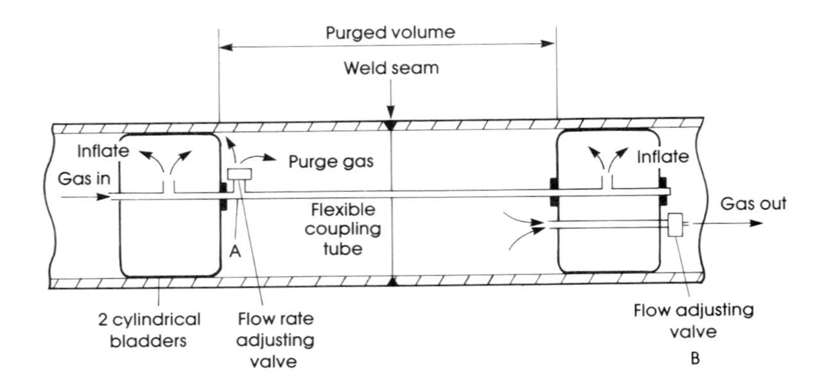

13.4 Inflatable dams or bladders for back purging.

The purpose of back purging is to minimise the amount of free oxygen remaining in the shielding gas. Ensure that your gas flow rate is adequate and that the pre-purging period is of sufficient duration.

Free oxygen meters

These are readily available as proprietary items and should be used in the outgoing gas line from the purge system. Try to achieve an oxygen content of *less* than one per cent for best results.

Watercoolers

In heavy and continuous TIG welding the torch can become extremely hot and uncomfortable for the operator. Usually torches with a capacity of 125 A or less get adequate cooling from the flow of shielding gas passing through the torch to the arc area, and these seldom need any extra cooling. Torches with a capacity of 125 A upwards need some extra assistance by use of a watercooler which circulates water around the internal torch body, usually entering through a single tube and returning to the cooler via a sealed tube which has the incoming power cable concentric with the tube, with a space between the cable and the tube wall. See also under cables, chapter 3.

Watercoolers are generally small, light and compact as they only need to provide a flow of water of about 7 l (1.5 gallons) per minute. They are usually of the recirculating type with an anti-rust header tank containing about 14 l (3 gallons). It is good practice to add a small quantity of anti-freeze/rust inhibitor to the top-up tank when in continuous use in extreme temperature conditions. It is useful to note that a system employing small orbital or bore weld heads needs deionised water in the cooling system. This not only protects the system but in certain cases can also help with arc strike.

Tungsten electrode grinders

Electrodes for hand welding can be sharpened manually quite successfully, using a bench grinder fitted with suitable grinding wheels as already mentioned in chapter 3. Also in chapter 3 the optimum point geometry for machine welding was discussed and illustrated.

For fine and fully specified machine welding to be successfully executed on a consistent basis, all parameters including the tungsten tip shape must be exactly repeated each time a weld is made. To achieve repeatable points at each sharpening a dedicated mechanical grinder is a must. There are many on the world market with an enormous price differential so the choice is up to you and will probably be determined by economics, *i.e.* how many resharpenings will be required and to what accuracy.

If budget allows, the author recommends any well made machine which rotates the tungsten slowly whilst the tip is being ground at high speed, Fig. 13.5.

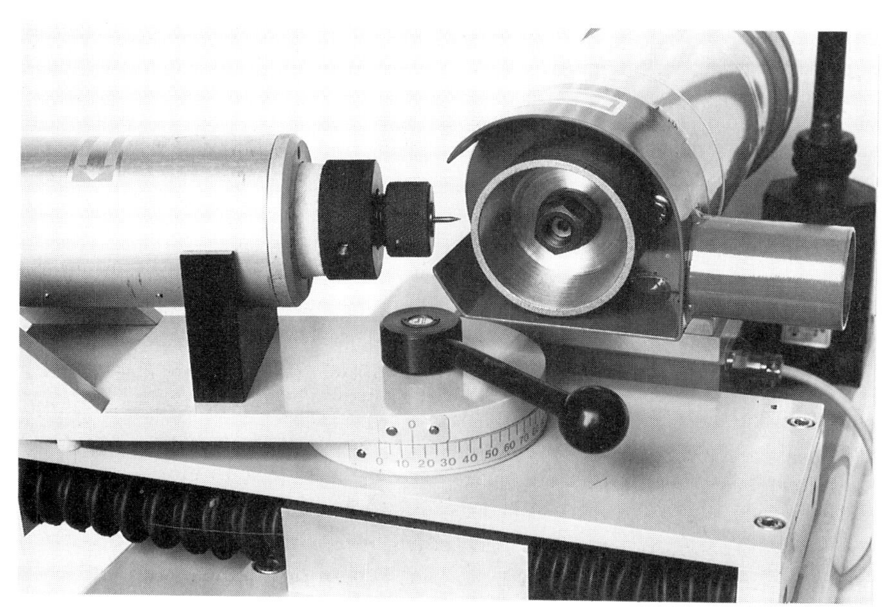

13.5 Precision bench mounted tungsten electrode grinder.

These are, however, rather expensive, so failing this if you cannot afford one of this type ensure that the grinder you choose leaves the tips finished as in chapter 3. Many extra accessories are available for all machines.

For very fine long points diamond wheels are a great advantage, but again expensive, so use them with care.

In passing it should be noted that electrode point sharpening *liquids* are available. The method here is to heat up the tungsten tip to a cherry red by shorting it out against the ground (earth) clamp and the tip is then immersed in the liquid, repeating this until the desired point is formed. This operation is claimed to take about one minute. However, the author has no experience with these liquids but feels that they could not produce accurate, consistent points.

Arc length control systems

Sometimes it is not possible for a weld seam in a circular component to be fully concentric or for a longitudinal seam face to be level. It then becomes necessary to try and keep the arc gap consistent to equalise heat input and thus ensure the quality and appearance of the finished weld. Whilst lateral traverse or rotational speed can be kept to fine limits using tachogenerator feedback (TGF) on the motor drive control system which moves the torch or component, the most usual method of arc length control (ALC) is based on the fact that arc voltage is proportional to arc gap. The TIG welding process is most suitable for this type of equipment.

In these systems arc length is maintained by constantly measuring arc voltage. When a variation occurs against a set datum voltage for a particular gap, an electronic signal is referred to a stepper motor driven slide on which the torch is mounted to adjust the electrode to work gap accordingly and keep the gap constant, Fig. 13.6.

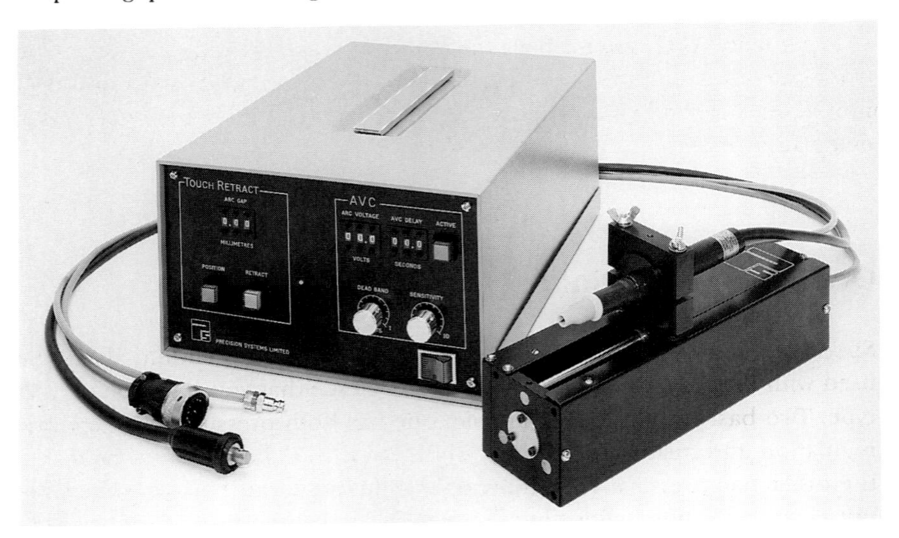

13.6 Touch-retract arc length controller with torch adjusting slide.

Several other methods using lasers or acoustics *i.e.* listening to the sound of the arc, have been used but voltage measurement remains the most successful and most economical method of arc length control.

Voltage measurement ALC systems often incorporate a touch-retract gap setting system. This can be an advantage in an automatic TIG system to give a sequence as follows:

1 An initiate button is pressed with electrode in the datum position;
2 Electrode advances towards the component to be welded and touches it;
3 A low voltage circuit is completed and the electrode instantly retracts;
4 When a preset gap is reached using, say, a stepper motor or encoder to count exactly the revolutions of a leadscrew, the retraction ceases;
5 A signal is then sent to the power source to initiate the weld sequence;
6 Sequence goes through pre-purge to arc strike;
7 After upslope (if any) and a preset delay period, arc length control if selected will commence. Fixture movement starts;
8 Welding continues for the required arc on time, with the ALC circuit sensing the changing arc gap and adjusting it accordingly;
9 Weld time finishes and arc is extinguished. ALC off;
10 After downslope, fixture motion ceases. Post-purge continues;
11 After post-purge the electrode returns to datum;
12 Sequence over – a new component is positioned and sequence repeated from 1.

If components are accurate and level or concentric the ALC function may not be needed. In such cases a touch-retract only system will suffice, simply to set the arc gap accurately for each sequence regardless of wear at the electrode tip.

The touch down period is negligibly short before retraction takes place so even very fine electrode points are not damaged.

Top quality ALC systems have the ability to react to changes in the arc to work gap of about 3 mm in 1 second, positional resolution of around 0.01 mm (0.0004 in) and voltage setting to 0.1 V, plus sensitivity and ALC on/off delay adjustments.

Gas cylinder regulators

Special regulators are available specifically for argon and should always be used with that gas. Check with your dealer to see that you have the correct type. Two basic models are available: one has both pressure and flow rate regulation and meters, to be used when gas is piped directly to the torch; the other has pressure regulation only. The latter can be used when the power source is fitted with a gas flow regulator and should be set to around 2.4 – 3.4 bar (35 – 50 PSI) for best results.

The gas output tube connection for argon usually has a right hand (RH) thread. To avoid accidents, regulators for argon/hydrogen mixes had a left hand (LH) thread on the gas line connector. For safety, ensure that a shut off key is attached to the cylinder by chain or cord and readily accessible so that gas flow can be quickly terminated in the event of emergency.

Arc viewing

Equipment to perform this function falls mainly into two categories:

1 Fibre optics and a viewing/magnifying screen, Fig. 13.7.
2 Closed circuit television (CCTV) systems, Fig. 13.8.

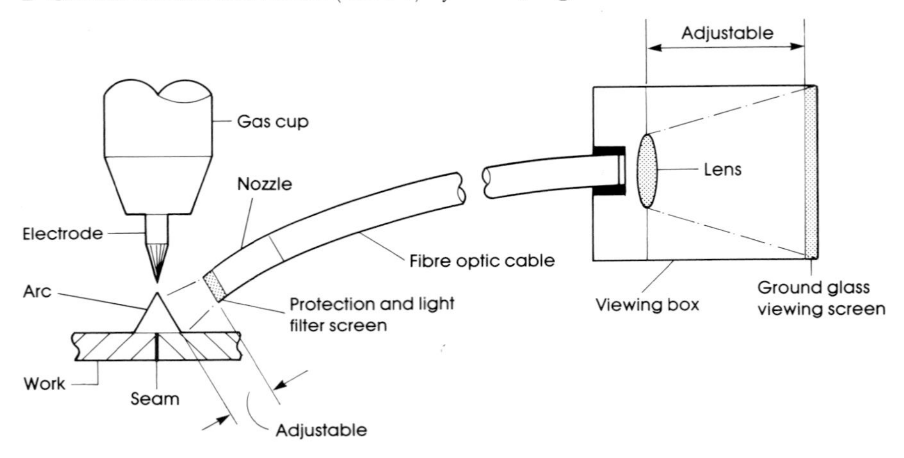

13.7 Fibre optic arc viewing system.

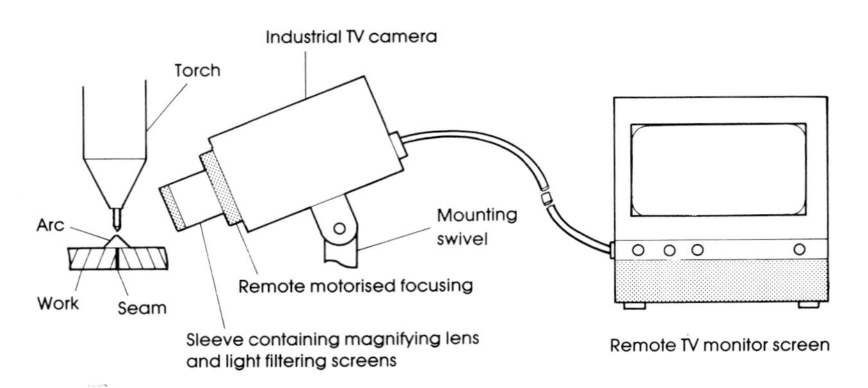

13.8 Closed circuit television arc viewing system.

Item 1 has two drawbacks, one that the operator's viewing screen cannot be very far away from the arc because of light attenuation in the fibre optics bundle and degradation of the pick-up tip at the point nearest the arc. Nevertheless these systems are efficient if regular cleaning of the probe tip takes place, and they are not too expensive.

Item 2 is rapidly improving, mainly through continuing miniaturisation of TV cameras. The operator can watch the arc from a considerable distance and adjust the focus to suit. The camera lens assembly can also be situated well back from the arc and protected from its effects.

Arc starting devices

The early method of starting an arc when TIG was still used exclusively for hand welding was to scratch the electrode on the work then rapidly withdraw it to make an arc before the electrode fused to the work. Operatives obtained a high degree of skill in doing this but as TIG progressed to be used on expensive metals and critical joints, it became obvious that an open arc start must be performed to avoid problems with tungsten inclusions in the weld. One method still used today is to use capacitor banks or a coil to produce a high frequency spark which rapidly ionises the gas in the arc gap and allows the welding current to use this as a bridge for arc initiation. HF, as it is called, is messy and interferes with ancillary equipment such as DC motors either through mains borne spikes or RF radiation. Use of computers in close proximity also made it imperative to replace this as much as possible, so modern power sources use electronically produced impulse strike methods, each with its own advantages and disadvantages. Check on this before buying a power source. Some old scratch start equipment is still in use, a tribute to its strong construction.

Hot wire feeders

The extra heat needed to melt a wire infill slows metal deposition, so methods of preheating the wire as it enters the weld pool have been devised. Some units can be added to existing systems and are commercially available from specialist suppliers who, in some cases, claim that they have deposition rates as good as in MIG welding (see chapter 7).

Cladding benefits from this technique where it reduces the amount of preheating of the base metal. The process is also particularly suitable for heavy plate welding for filler and capping runs. Most suitable for mechanical TIG welding systems, check carefully as to the economics before purchase.

Penetration control

Systems are available which view the weld underbead using fibre optics and/or a CCTV camera. Light emerging from the underbead is measured and corresponding feedback referred to a computerised control unit. This in its turn adjusts the welding current parameters at the power source, ensuring a consistent underbead size and width and thus the required weld penetration. These units are expensive but, if the amount of penetration is critical, can be invaluable.

Chapter 14

Safety

TIG welding properly carried out is not particularly hazardous. However, as in all engineering processes, care should be taken. This chapter briefly covers the main areas where attention to safety is vital, and is based mainly on UK standard practice. Other countries will have equivalents.

Eye protection

TIG welding does not usually produce much spatter except with AC where it is more pronounced, and commercial eyeshields and screens take care of any that may arise. Spatter is often the result of unclean or damp workpieces but some alloys and coated materials also cause problems in this respect.

What *is* essential, however, is that all eye screens should be of sufficient optical density as the TIG arc is extremely intense. Screens are graded for density and it is the author's opinion that the screen to use should be of the maximum possible opacity consistent with adequate arc and work viewing.

Grading of density in the UK is often done by marking screens, particularly replacement screens for helmets, with an EW (electric welding) rating from 1 – 15, grade 15 being the most opaque. Nothing below EW8 should be used for TIG welding, a good grade being EW10 which is quite dense. This can make it difficult to see before the arc lights up the weld area. Peripheral viewing of arcs by other personnel can be prevented by hanging suitable semi-opaque curtains around the work areas. There are on the market certain visor screens and goggles which are clear until the arc is struck, similar to light reacting sunglasses but with a much more rapid response time. At the time of writing these screens do not appear to have met with any significant success but their makers claim that they are perfectly safe. Certainly it is a great advantage to be able to see the exact location for starting a weld before striking an arc but it is suggested that extensive trials of such reactive screens take place before comprehensive use is made of them.

In the end the choice of helmet or goggles, conventional or reactive screens will be made by the welder himself and will be of the type considered the most comfortable and convenient.

A flash, or arceye as it is called, can be very inconvenient and uncomfortable. This occurs after an arc is accidentally viewed without a safety screen. Usually a day or two of rest and wearing dark glasses produce a cure. Eyedrops can help but these should *only* be used on professional medical advice. Never ignore a flash, take some action.

Cleaning solvents

Solvents containing trichlorethene (formerly trichlorethylene), perchlorethene (perchlorethylene) *etc* must only be used in a well ventilated area and allowed to dry off before welding. These substances can produce phosgene, a very toxic gas, in the heat of the arc. Trichloroethane is a safer solvent.

Safe handling, storage and use of compressed gas cylinders

The following practices are recommended for safe handling, storage and use of high pressure gaseous and liquefied compressed gases. Additional precautions may be necessary depending upon the category to which the gas belongs (corrosive, toxic, flammable, pyrophoric, oxidant, radioactive or inert), the individual properties of the gas and the process in which it is used.

GENERAL

- Only experienced and properly instructed persons should handle compressed gases.
- Observe all regulations and local requirements regarding the carriage and storage of cylinders.
- Do not remove or deface labels provided for identification of cylinder contents.
- Ascertain the identity of the gas before using or transporting it.

- Know and understand the properties and hazards associated with each gas before using it.
- Before using or transporting compressed gases, establish plans to cover any emergency that might arise.
- When doubt exists as to the correct handling procedure for a particular gas, contact your supplier.
- If you own your cylinders you must be aware of and discharge your statutory obligations with regard to maintenance and testing. These regulations are constantly under extensive review.

HANDLING AND USE

- Wear stout gloves.
- Never lift a cylinder by the cap or guard, unless the supplier states it is designed for that purpose.
- Use a cylinder trolley or other suitable trolley for transporting cylinders even for a short distance.
- Leave valve protection caps in place until the cylinder has been secured against either a wall or bench, or placed in a cylinder stand or trolley and is ready for use.
- Check for gas leaks using approved leak detection solution.
- Ascertain that an adequate supply of water is available for first aid fire fighting in the event of leakage.
- Use suitable pressure regulating devices on all cylinders when the gas is being delivered to systems with a lower pressure rating than that of the cylinder.
- Before connecting the cylinder for use ensure that backfeed from the system into the cylinder is prevented.
- Before connecting a cylinder check the complete gas system for suitability, particularly for pressure rating and materials.
- Never permit liquefied gas to become trapped in parts of the system as this may result in hydraulic rupture.
- Ascertain that all electrical systems in the area are suitable for service with each gas.
- Never use direct flame or electrical heating devices to raise the pressure of a cylinder. Cylinders should not be subjected to temperatures above 45 C.
- Never recompress a gas or a gas mixture from a cylinder without consulting the supplier.
- Never attempt to transfer gases from one cylinder to another.
- Do not attempt to increase liquid drawoff rate by pressurising the cylinder without first checking with the supplier.
- Do not use cylinders as rollers or supports or for any other purpose than to contain the gas as supplied.
- Never permit oil, grease or other readily combustible substances to

come into contact with valves of cylinders containing oxygen.

- Keep cylinder valve outlets clean and free from contaminants, particularly oil and water.
- Do not subject cylinders to abnormal mechanical shocks which may cause damage to their valves or safety devices.
- Never attempt to repair or modify cylinder valves or safety relief devices. Damaged valves should be reported immediately to the supplier.
- Close the cylinder valve whenever gas is not required even if the cylinder is still connected to equipment.

STORAGE

- Cylinders should preferably be stored in a purpose built compound which should be well ventilated, preferably in the open air.
- Store cylinders in a location free from fire risk and away from sources of heat and ignition.
- The cylinder storage compound should be kept clear, access should be clearly marked as a cylinder store and appropriate hazard warning signs displayed, *e.g.* flammable, oxidant, compressed gas, *etc.*
- Smoking and the use of naked flames either inside or in the vicinity of the cylinder storage area should be prohibited.
- Cylinders should be stored in the vertical position and properly secured to prevent toppling. The cylinder valves should be tightly closed and, where appropriate, valve outlets capped or plugged. Cylinder valve guards or caps should be in place and properly secured.
- Protect cylinders stored in the open against rusting and extremes of weather. It is advisable to stand cylinders on open galvanized steel gridwork to reduce corrosion of the cylinder base.
- Store full and empty cylinders separately and arrange full cylinders so that the oldest stock is used first.
- Gas cylinders should be segregated in the storage area according to the various categories, *i.e.* toxic, flammable, *etc.*
- Cylinders containing oxygen and oxidants should be separated from flammable gases by a minimum distance of 6 m (20 ft) or, alternatively, a fire resistant partition.
- Do not mix cylinders in the full cylinder store. Store full cylinders of different gases separately, each in a well marked place.
- The amounts of flammable or toxic gases in storage should be kept to a minimum.
- Cylinders containing flammable gases should be stored away from other combustible materials.
- Cylinders held in storage should be periodically checked for general condition and leakage.

CARRIAGE

- Make sure that the driver who carries cylinders in a vehicle, particularly flammable and toxic gas cylinders, has been properly instructed in the method of handling and loading cylinders, in dealing with any emergency and carries the required information.
- Carry cylinders in open vehicles if possible. A closed vehicle may be used to carry small quantities of cylinders if well ventilated.
- Ensure cylinders are properly secured on the vehicle and that propane cylinders are always kept upright during carriage.

Further reading

Handbook of compressed gases, Compressed Gas Association Inc, Reinhold, 1981.

Patty, F A editor, 'Industrial hygiene and toxicology', 2nd edition vol 2, John Wiley & Sons, 1962.

Gas Data Book, Matheson Gas Products, 1971.

British Compressed Gas Association, CP9-Code of Practice 'The safe filling, handling, storage and distribution of gases in transportable containers', 1982.

'Safe under pressure', BOC Ltd.

Gas encyclopaedia, L'Air Liquide, Elsevier, 1976.

The Road Traffic (carriage of dangerous substances in packages, *etc.*) Regulations 1986, SI.1986, No. 1951 and supporting code of practice.

Notes on argon hazards and basic safety precautions

FIRE AND EXPLOSION HAZARDS

Neither gaseous nor liquid argon is flammable and do not in themselves constitute a fire or explosion risk. However, they are normally stored under pressure and the storage vessels, whether gas cylinders or liquid tanks, should not be located in areas where there is a high risk of fire or where they may normally be exposed to excessive heat. Containers of compressed gaseous argon may rupture violently if overheated as a result of exposure to fire.

Oil lubricated compressors operating continually on argon service for a prolonged period should not be switched to air service without thorough cleaning, otherwise there is a danger that unoxidised pyrophoric deposits which may have formed in the machine will explode violently on contact with compressed air.

Asphyxia

Argon, although non-toxic can constitute an asphyxiation hazard through displacement of oxygen in the atmosphere. The potential for this type of hazard is significant because of the widespread use of argon in industry. Neither argon nor oxygen depletion is detectable by the normal human senses. Unless adequate precautions are taken persons can be exposed to oxygen deficient atmospheres if they enter equipment or areas which have contained argon or in which argon has been used.

Symptoms of oxygen deprivation, *e.g.* increased pulse and rate of breathing, fatigue and abnormal perceptions or responses, may be apparent at an oxygen concentration of 16%. *Breathing a pure argon atmosphere will produce immediate loss of consciousness and almost immediate death.*

PRECAUTIONS

Operation and maintenance

It is essential that operations involving use of gaseous or liquid argon, particularly in large quantities, are conducted in well ventilated areas to prevent formation of oxygen deficient atmospheres.

Ideally, argon should be vented into the open air well away from areas frequented by personnel. Argon should *never* be released or vented into enclosed areas or buildings where the ventilation is inadequate. Both cold argon vapour and gaseous argon at ambient temperature are denser than air and can accumulate in low lying areas such as pits and trenches.

EMERGENCIES

In the event of accident or emergency the instructions below should be implemented without delay.

Asphyxiation

Persons showing symptoms of oxygen deprivation should be moved immediately to a normal atmosphere. Persons who are unconscious or not breathing must receive immediate first aid. Medical assistance should be summoned without delay. First aid measures include inspection of the victim's airway for obstruction, artificial respiration and simultaneous administration of oxygen. The victim should be kept warm and resting.

Further reading

Cryogenics Safety Manual 'A guide to good practice' published by The British Cryogenics Council, London SW1.

Prevention of accidents arising from enrichment or deficiency of oxygen in the atmosphere 'Document 8/76E Industrial Gases Committee of CP1 Paris 1976'.

Fume extraction

For most TIG welding all that is needed is good ventilation to avoid gas build-up. Argon is denser than air and tends to concentrate in low lying areas and is very seldom a problem. Just an open door in the welding area is usually all that is needed, ensuring of course that no draughts can disturb the gas shield around the arc.

TIG welding produces very little smoke but does form ozone, particularly when welding aluminium and stainless steels, which has a characteristic smell associated with TIG welding. Ozone should *not* be extensively inhaled although exposure to the smell usually causes no harm.

HEALTH RISKS OF WELDING FUME

(from HSE (Health and Safety Executive) Guidance Notes EH54 and EH55 – UK only.)

Welding fume is a mixture of airborne gases and fine particles which if inhaled or swallowed may be a health risk. The degree of risk depends on:

- The composition of the fume;
- The concentration of the fume;
- The duration of exposure.

The main health effects are:

- Irritation of the respiratory tract. Gases or fine particles of fume can cause dryness of the throat, tickling, coughing, tightness of the chest and difficulty in breathing.
- Metal fume fever. Inhaling many freshly formed metallic oxides, such as those of zinc, cadmium, copper, *etc* may lead to acute influenza-like illness termed metal fume fever. With the exception of exposure to cadmium fume serious complications are rare. The commonest cause of metal fume fever is welding galvanized steel.
- Systemic poisoning. Systemic poisoning can result from inhaling or swallowing substances contained in welding fume such as fluorides, hexavalent chromium, lead, barium and cadmium. The presence of these substances in the fume depends upon the welding process being used and the material being welded.
- Long term or chronic effects. Inhaling welding fumes can lead to siderosis or pneumonoconiosis. A subject of current concern is whether welders have an increased risk of developing respiratory cancer, as certain constituents of some welding fumes, *e.g.* hexavalent chromium and nickel, may be carcinogenic.

Document EH 54 or a related publication of other countries' safety councils should be obtained and studied carefully. All such councils have set limits to exposure times and levels, along with recommended limits of particle inhalation. As TIG introduces filler metal into an already molten weld pool there is no transfer of metal particles directly through the arc, so escape of metal particles is minimal. Study the documents carefully and equip your welding area accordingly.

Operators should not, if possible, position their heads directly over the arc unless wearing an extractor mask. For normal purposes the helmet deflects fume and for extra safety a simple cloth mouth and nose face mask will suffice. Very low current precision TIG welding needs no mask at all, but ensure good ventilation at all times. For further details, and a list of relevant regulations, consult COSHH (Control of Substances Hazardous to Health) and any applicable British and European Standards, *etc*. (Also in the UK consult BS6691.) In other countries obtain the relevant literature from the official authority. Further data can usually be obtained from manufacturers of welding consumables.

14.1 Fume extractor for manual TIG welding.

METHODS OF FUME EXTRACTION

Figure 14.1 shows an extractor duct for MIG welding, the requirements for TIG are similar but not as demanding. The main point is that suction should not be too fierce or the inert gas shield will be reduced, and do not forget, gases cost money.

Collection of dust and metal particles from the fume is useful but does not present too much of a problem for TIG welding. However, fit the best extraction you can afford consistent with the work being done.

Chapter 15

Future developments for TIG welding

Over the whole history of TIG welding the working end, *i.e.* the torch, electrode and gas, have remained much the same since its inception. Torches have become more versatile and lighter, reasonable quality electrodes are offered in a large range of sizes and material content and gases are available in an even greater variety of purity and mixes. So, not much improvement left to make? Well, human ingenuity is always devoted to producing cheaper and better technical items, so maybe we will have to wait and see what improvements, if any, are to be made in those areas.

Where improvements *will* be made is undoubtedly in equipment, *i.e.* power sources, control systems, monitoring, arc viewing and data recording, *etc.* Systems already exist for remote control of arc and electrode position, along with closed circuit TV viewing in inaccessible areas. Mechanical handling and arc positioning are already used in 'hot' nuclear areas with great success and the author has seen many ingenious installations of this type.

Electronics will be improved to a very high standard to provide smaller, lighter and more efficient power sources. TIG, of all the welding processes, is extremely well suited to computer control and most future equipment will be upgraded to allow for computers to be fitted on an add-on basis.

In the author's opinion there are some areas where further attention could be paid to improving equipment and lowering the cost. Arc length control, seam following and arc viewing equipment are capable of further development and would be more widely used if they were not as costly as at present. A considerable amount of electronic components are used and these are getting cheaper and easier to obtain. Many power source control and regulation circuits are now solid state and more will become so.

Regarding control of automated mechanical systems, this is already widely carried out by pneumatic logic so further detailed development to allow these circuits to be integrated with the power source and manipulation systems is necessary.

Arc starting devices must be developed always to be computer friendly leading, the author hopes, to the eventual abandonment of the use of primitive HF with all its problems.

Some research must be done by manufacturers to produce better and more consistent electrodes. TIG alone poses the problem where extremely expensive welding systems depend on an item costing a few tens of pence. The complaint here is not about cost but quality. Perhaps there is an element other than thorium, zirconium, cerium or lanthanum that will have all their virtues and none of their drawbacks.

Any developments must be seen to be advantageous and progressive, not as is often the case, change for change's sake. As the current saying goes 'We have the technology', let us use it for real improvement. If we do this, TIG welding will have a bright and commercially viable future.

Finally the author would be pleased to hear readers' comments on the material in this book. Nobody knows it all and argument, disagreement, discussion and constructive criticism should bring progress for the good of industry in general.

Index